SİLVER

The Earth series traces the historical significance and cultural history of natural phenomena. Written by experts who are passionate about their subject, titles in the series bring together science, art, literature, mythology, religion and popular culture, exploring and explaining the planet we inhabit in new and exciting ways.

Series editor: Daniel Allen

In the same series

Air Peter Adey
Cave Ralph Crane and Lisa Fletcher
Clouds Richard Hamblyn
Desert Roslynn D. Haynes
Earthquake Andrew Robinson
Fire Stephen J. Pyne
Flood John Withington
Gold Rebecca Zorach
    and Michael W. Phillips Jr
Islands Stephen A. Royle
Lightning Derek M. Elsom

Meteorite Maria Golia
Moon Edgar Williams
Mountain Veronica della Dora
Silver Lindsay Shen
South Pole Elizabeth Leane
Storm John Withington
Tsunami Richard Hamblyn
Volcano James Hamilton
Water Veronica Strang
Waterfall Brian J. Hudson

Silver

Lindsay Shen

REAKTION BOOKS

For my parents, Ewan and Audrey Macbeth, in gratitude

Published by Reaktion Books Ltd
Unit 32, Waterside
44–48 Wharf Road
London N1 7UX, UK
www.reaktionbooks.co.uk

First published 2017

Printed and bound in China by 1010 Printing International Ltd

A catalogue record for this book is available from the British Library

ISBN 978 1 78023 756 5

CONTENTS

Preface: A Silver Object Lesson

Every cloud has a silver lining. And silver, like clouds, is seemingly everywhere. The aim of this book is to consider something that we have become so familiar with that we easily take it for granted. It is the stuff of our grandfathers' coin collections and the hidden contents of our grandmothers' cabinet drawers; it is in old photographs, our mobile phones and computer keyboards, our teeth, water, swimming pools, socks and bandages; it is what our New Age neighbours swallow as homeopathic medicine, and what investment gurus urge us to buy instead of equities; it is the banqueting dish offered at half a million pounds through the auction house, and the necklace on sale for £14.99 at the local accessories shop; it is in our language, our sayings and ideas (reports of 'silver bullets' that will cure our problems surface in the media almost daily). Increasingly, with the development of nanotechnology, it is also in our bodies and those environments that have never been a natural home for silver. As an element, it is one of the building blocks of the earth, and like other building blocks such as oxygen, hydrogen, lead, sodium or iron, it simply exists.

But how? How does one silver object – let us take a water flacon by the British silversmith Charles Hall – come into being? At 24 cm (9.5 in.) high, the flacon is a modestly scaled object designed for holding and pouring water. The hallmark on its base provides us with a potted history of its making. The sponsor's mark 'cfh' is the initials of its creator. The letter 'n' is the date mark, indicating that it was hallmarked in 2012. The leopard's

Charles Hall, water flacon, 2012, silver.

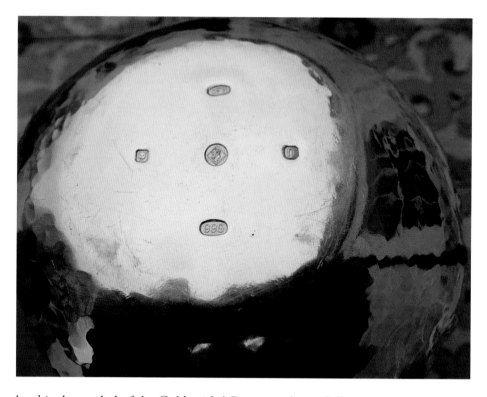

head is the symbol of the Goldsmiths' Company Assay Office in London, where it was sent to authenticate the purity of the silver. The mark '999' is the millesimal fineness mark, which tells us it is 999 parts out of 1,000 pure silver. This is the standard we call fine silver, though most of the silver that we are familiar with is the lower purity sterling silver – 925 – meaning 92.5 per cent pure. The mark in the centre, depicting the Queen's head in a diamond, is a special commemorative mark to celebrate Queen Elizabeth II's Diamond Jubilee. It was handmade in Cornwall, but the silver did not come from a Cornish mine; there were silver (and tin, lead, zinc and copper) mines in Cornwall, but these have long since been worked out. Charles Hall sourced his silver in Italy, but his supplier drew on several sources. It did not come out of the ground ready to be worked. Unlike gold, little pure 'native' silver exists in the earth. The material for this flacon had to be extensively refined before it reached Hall.

Hallmark on Charles Hall's water flacon.

This book first asks the basic question, where does silver come from before it reaches the silversmith's workshop, medical laboratory or electronics factory? The silver for Hall's flacon made a journey that can only be measured in geological time. Ore genesis (the forming of ores) takes millions of years. But, projecting beyond the formation of the earth, how did that process begin? Why are silver deposits found in certain places? What sort of geology allows silver to come into being? Nowadays mining exploration relies on geologists, geoscientists and engineers to determine the viability of mining in a promising location. But how were silver deposits discovered thousands of years ago? And what beliefs and mythologies grew up around the revelation of such a precious metal, the effort and expertise required to prise it from the earth and the almost mystical process of refining it?

An ancient idea held by many cultures is that of the miner collaborating with nature in 'birthing' the precious metal. The reverse of this is the smith, through skill and ingenuity refashioning nature into art. Hall's flacon, with its beautifully rounded shape that asks to be handled, was raised from a flat sheet of silver. Little wonder that the metalsmith has long been thought a magician. Raising is one of the oldest silver-working techniques we know about, and involves the smith hammering cold silver, which has a natural softness, over a form to create a dish or hollow vessel. A complex shape like the flacon, with its narrow neck, is more easily made from fine silver, which is especially soft. The greater the percentage of another material (usually copper), the harder the silver. It is not unusual for a smith to craft new tools for a project, such as the one Hall made to shape the specific curve of the vessel.

The flacon's surface is textured with very visible hammer marks that change size and shape as they move up towards the rim. The hammering is functional in that it hardens the soft, pure metal, but it also creates shallow concavities that enhance the play of light over the surface of the earth's most reflective metal. It is a surface that pools and flows and glistens – like the water inside. The hammer marks were made by the hand of one particular silversmith, but belong to a craft stretching back

millennia. The reverse of the question 'how is silver formed in nature?' is 'how do humans form silver?' How do silversmiths like Hall work with silver's natural properties, such as malleability and shine, to create objects of use, beauty, adornment and even magic? Silver has been cast into objects that even in their solid form seem fluid and living. It has been manipulated with such craft that over the millennia, smiths have conjured prancing horses, dancers, deities, flowers and landscapes from flat surfaces. It has been 'drawn on' with engraving tools, drawn into wire, woven into sumptuous textiles and even crocheted.

Charles Hall, silver-working tool for making his water flacon.

The hallmark on Hall's flacon belongs to a British institution of marking silver that stretches back 700 years. But the practice of having some official authentication of the purity of silver reaches back much further. This is because silver, like gold, has been a way to store wealth. If the silver was debased, the value eroded. The next part of this book looks at silver as wealth, and the transformations of nations, political systems and ideas through the flow of silver as currency. From the advent of silver mining in Anatolia and surrounding regions over 5,000 years ago, silver has sustained economies, fostered trade and bankrolled the building of empires. Without the silver mines at Laurion and the wealth these injected into Athens, we might have inherited an entirely different culture from the classical world. Without Spanish control of the phenomenally productive silver mines in modern-day Bolivia, the economic development of Europe might have evolved differently. Possibly Asia's trading relationships with the West, and even political institutions, might have taken other forms. Those small marks on Hall's flacon that point to one person, one year, one place, one measure of purity, in fact orientate it outwards into the world.

For most of silver's history, the idea that it has intrinsic value and can be traded for other things shaped demand. But what happens as our ideas about financial tools change? As silver has become less useful as currency, its value has increased in the industrial sectors, and new flows of demand lead in multiple directions. For instance, when Charles Hall exhibited his flacon in the Victoria and Albert Museum in 2012, its expressed purpose was 'to convey water elegantly, enhanced and purified through fine silver'. Silver has been valued as a purifying agent since ancient times. The Greek historian Herodotus noted that the Persian king Cyrus the Great stored boiled river water in silver during military campaigns – a recorded instance of a general knowledge of what we now understand as its antimicrobial properties. Chemistry helps explain why; intriguing recent research posits a 'zombies effect', in which bacteria killed by silver then massacre their healthy neighbours. Before germ theory, which gained support from the mid-1800s, it was an observable principle that silver helped preserve freshness. Now silver is applied to the insides of water tanks and to water filtration systems. It was used as a germicide until antibiotics became available after the Second World War, but with the increasing resistance of bacteria, new means of incorporating silver into antimicrobial therapies are being developed all the time. Another of silver's properties is its high conductivity, which has led to multiple uses in the electronics industry. When we think of silver, we might visualize jewellery or family heirlooms, but today most demand for silver is from industrial sectors such as energy and healthcare.

We might instead think of antibacterial socks rather than knives and forks.

Despite our knowledge of the salutary benefits, we do not generally serve water from silver pitchers, or eat with silver cutlery from silver plates. Silver remains a precious metal, and like gold has long been used to advertise its owner's status. The final part of this book moves into the realm of symbols. Today we might measure our worth by the number of followers we have on our social media accounts, but such abstract validation is fairly new. For much of history, we have signalled our status through physical belongings that can be seen and envied by others. Because of their rarity and visual properties, like sheen, that we tend to find beautiful, precious metals such as silver have long been admirable assets. A silver platter engraved with a family coat of arms could, in times of need, be hastily melted down into bullion, but as an object ostentatiously displayed on the dinner table, it was a mark of success beyond mere buying power.

Silver purifies, but it is also symbolically pure. For this reason, it can vanquish werewolves and vampires, deflect malevolent spirits and ward off the evil eye. Worn as an amulet, it can protect the vulnerable, such as infants, or those embarking on journeys, including that long and perilous one to the afterlife. Since we first scraped silver from the surface of the earth, we have imbued it with realms of meanings that take us beyond our planet and its silver-lined clouds, far into the worlds of imagination. Matter from stars, it leads us back there.

1   The Nature of Silver

Silver is a siren. It is the earth's most shiny, reflective metal, spinning light into our eyes like a flipped silver dollar, or drachma, or peso, blinding us to its corruptive potential and hurling men and empires (and vampires) to ruin. We have coveted it – sometimes over its rarer and costlier cousin, gold – since we first learned how to extract it from the earth. It has been awarded to our victors: at the first modern Olympic Games in Athens, in 1896, the winners were crowned with an olive wreath and awarded a silver medal; the 'gold' medals of subsequent first-place winners have almost always been gold-plated silver.[1] Silver brings out our most generous instincts – as, for example, when we give it as a gift to celebrate a marriage, birth or special achievement – and our very worst. A highly publicized wrangle over ownership of the 'Sevso Treasure', a spectacular hoard of exquisitely crafted late Roman silver, involved three alleged murders before some of it was repatriated by the Hungarian government in 2014.

All this, of course, has more to do with human nature than the nature of silver. But what is it about silver that incites these sentiments and actions? After all, just like gold, diamonds, emeralds – all of the precious metals and gems for which we profess love and wage war – silver started out as space debris. The question of why we value some things over others has exercised philosophers for millennia. As the fourth-century Christian leader John Chrysostom so profoundly enquired of his increasingly materialistic flock, 'Are gold and silver beautiful? But consider that they were and still are dirt and ashes.'[2] Chrysostom was urging the errant Christians of late Antiquity to accumulate spiritual rather than material capital, but his reference to filth was double-edged. While coveting earthly treasure might be degrading, it is also nonsensical – if everything has the same origin and comes to the same end, why value silver over, say, zinc?

Silver structure

Perhaps if we take a closer look than Chrysostom was able to, we might understand silver's allure a little better. What we currently understand about silver's structure and behaviour in relation to other substances can help explain the properties we prize, though knowledge shifts constantly, especially of a material so diversely useful to contemporary life. Silver has the atomic number 47, with 47 positively charged protons in its nucleus, and 47 negatively charged electrons around the nucleus. These atoms pack together to form crystals in an orderly system we call metallic bonding. When metal atoms come together, most of the electrons are held tight by the positively charged nuclei, but some detach and join a 'sea of electrons' – a fluid environment for the flow of electrons that is an electrical current. Think of a riverbed that throws a few obstacles (rocks, roots, twists or tributaries) into the course of water. One of the reasons we value silver today is that, of all metals, it is the most effective conductor of electricity, and is used in high-performance electronics and electrical systems. The structure of silver creates an environment most conducive to this uninterrupted flow, though for reasons of cost, its close competitor copper is more widely used for everyday needs such as electrical wiring.

The malleability of silver – the extent to which it may be moulded to a different shape – means that it can be hammered into flat sheets, or formed into seamless round vessels. Some metals, such as titanium and steel, require heat to become malleable. Others, like gold and silver, are much softer and can be partially worked cold. These softer metals tend to have weaker metallic bonding, allowing the metal atoms to slide against each other under pressure. Similarly, weaker bonding results in gold and silver being ductile, characterized as being capable of being drawn into wires – a property exploited by silversmiths who have created delicate filigree ornament of thread-fine silver.

The symbol for silver on the periodic table of the elements is Ag, from the Latin *argentum*, whose root in both Latin and Greek means white or shining. Silver's perfect sheen comes from the fact that it is the most reflective of all metals. Over the light spectrum visible to humans (between about 400 and 700 nm), which we perceive as our gloriously multicoloured world, silver trumps all other metals in its constant reflectivity. Again, this has much to do with those free-flowing electrons. As waves of incoming light hit the electrons on the metallic surface, the agitated electrons create their own opposing field of energy that is pushed outwards in another wave – a reflection. This occurs

Reliquary cross,
Venice, *c.* 1750–1800,
silver.

in silver across all the wavelengths we can see, resulting in the reflection of clean and even light that we put to many uses. High-quality mirrors were historically coated with silver, and it is still used in scientific mirrors, precision optical equipment and telescopes.

Silver, along with gold, platinum and a small number of other metals, is classified as a noble metal, which means that it resists oxidation. Unlike iron, for example, it does not rust in contact with moisture-laden air or water. It does not readily react with oxygen in the air, though unlike gold, it does tarnish when exposed to sulphur compounds. The more heavily polluted the air, the quicker the build-up of black silver sulphide. A surprising consequence of living in the industrialized world is the need to frequently polish silver jewellery and tableware.

While family silver may not be much of a preoccupation of our contemporary world, advances in health care and infection control are. For centuries silver has been valued as a cleansing agent to purify water or disinfect wounds, but it is only recently that we have begun to understand just how silver destroys bacteria. Its antimicrobial properties are only activated through chemical reaction in silver's ionized form – that is, when the atoms, having lost an electron, become positively charged. Like all things organic, bacteria rely on enzymes to sustain life and flourish. It is these enzymes that silver ions attack and disable. In short order, the bacteria cells dehydrate, shrivel and die. An unexpected bonus is that the silver-impregnated dead cells 'infect' their healthy neighbours, leading to a wide-scale sustained massacre dubbed the 'zombies effect' by the scientists who recorded it.[3]

Origins

Among the breathtaking display of minerals and gemstones at the Freudenstein Castle in the historic mining town of Freiberg, Germany, is a specimen of silver. At a height of 15 cm (6 in.), it looks like an intricately twisted wire sculpture, or a sinuous underwater plant. Gorgeous specimens of pure or 'native' silver,

Native wire silver
with acanthite from
the Himmelsfurst
Mine, Brand-
Erbisdorf, Freiberg
District, Erzgebirge,
Saxony, Germany.

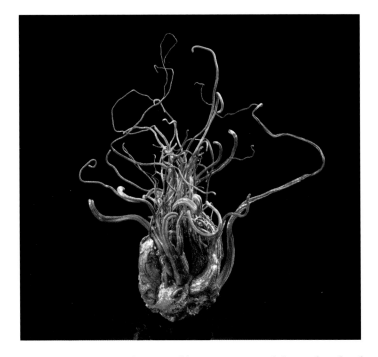

looking like cunningly pruned bonsai trees, or delicate fronds of
coral, belong to prestigious mineral collections throughout the
world. Readily collectible smaller versions are regularly offered
by dealers. Finger-length pieces, approximately 8 cm (3 in.),
might cost the same as a luxury car. As a centre of silver mining
for eight hundred years, the region around Freiberg was richly
productive. One mine produced a single specimen weighing 225
kg (500 lb) in 1857.[4]

Silver is widely dispersed around the globe, and specimens
of great substance and beauty have been found in places ranging
from Alaska to Australia to China – the last a new source for
small collectible pieces. Britain, too, had silver mines. The most
productive were in Cornwall, but Scotland had a small eighteenth-
century mine with rich native silver deposits at Alva, near Stirling,
where delicate tree-like crystals were retrieved in the 1980s.
Some of the most intricate and sought-after specimens originated in Kongsberg, Norway, which produced large masses of
lustrous wire, some a metre (3 ft) long. In terms of heft, however,

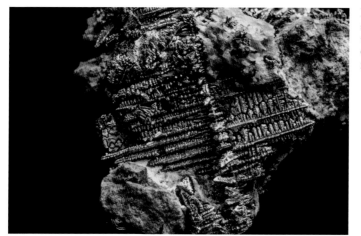

Dendrites of native silver, specimen from the Alva silver mine, Clackmannanshire, Scotland.

An entrance to the former Alva silver mine in Scotland.

these cannot compare with the slabs of silver discovered near Sonora, Mexico, or the sheets mined in Cobalt, Ontario, in the early twentieth century.[5]

Most mined silver does not look like this, though. Native silver is rare; it does not coil from clefts in exposed rock or gleam like gold in flecks on riverbeds. It is much more likely to exist in complicated mineral structures, locked into combinations with other elements such as lead, copper, chlorine or arsenic. We call these minerals 'ores' if they contain a valuable constituent (such as silver) that can be extracted for profit. Most silver-mining museums have samples of ores that to the untrained eye look unprepossessing, or even deceptive. For instance, a chunk of acanthite in quartz looks lumpen and sooty grey; proustite glints ruby-red. The story of how silver arrived in this form gives us a framework for understanding how we recognize it as silver, and retrieve it from the ground.

All earth's elements except the lightest, such as hydrogen and helium, were formed inside stars as they evolved, or at the point of their death. As stars burn through their fuel, lighter elements combine to create heavier ones, such as carbon, in a process known as fusion. Heavy metals like silver and gold, though, require an even more intense crucible in order to form – the death of the star itself, hurling elements into space as it explodes. The silver on our planet was forged from the supernovae of stars eight to nine times the mass of our sun. The stars that birthed gold were even larger, so although gold and silver are sometimes found together in nature, and although our culture often pairs them, their parents were quite separate.[6]

Fast forward – way forward – through the creation of the Earth and the tumultuous metamorphoses of its crust. One of the most common ways that silver concentrates near the surface where we can find it is through the action of hot fluids. These metals that were once space refuse dissolve in magma-heated water circulating deep within the earth's crust. This vast hydro-thermal plumbing system carries the mineral-charged solutions up into fissures and faults in rocks. As they cool, or react chemically with the rocks, the minerals are deposited. Sometimes they are exposed by erosion at the earth's surface; sometimes their discovery is more challenging. Silver starts out in space, but where it ends up is very much a matter of local conditions near the earth's surface, dependent on the pattern of faults and types of rocks.

Treasure hunt

Today throughout the world most silver is produced as a by-product of mining for other metals such as copper or lead. There are, however, substantial mines whose primary product is silver, and multinational mining companies that focus on discovery and extraction of silver. Discovering silver is both an art and a science, as no two deposits are exactly alike, and the veins, in relation to their surroundings, are vanishingly small. Although silver is more common than gold, its occurrence in the earth's crust is only a scant 0.07 parts per million.[7] The era of stumbling upon silver-rich outcrops is long gone.

Investigation into the economic viability of a specific piece of land for mining is termed exploration. Although large mining companies may conduct exploration with an in-house team of geologists and geophysicists, today a more common model is for small junior mining companies to conduct the primary research based on their preliminary identification of promising regions. Much of their work – mapping, sampling and drilling – is conducted on the ground. If a junior mining company succeeds in identifying a potential deposit, it typically sells its findings to a larger corporation which would take the project through development.[8] It is a high-risk game, as only a tiny number of prospects are ultimately developed as mines. In the preliminary stages, a common strategy is to return to areas with old mines and apply new technologies to identify additional, or more deeply buried deposits. According to a widely held mantra, 'when hunting elephants it's best to look in elephant country.' For instance, the American West's Comstock region, which was spectacularly bountiful in the nineteenth century, is still an area of interest today.

Landsat imagery
of Goldfield and
Tonopah in Nevada.

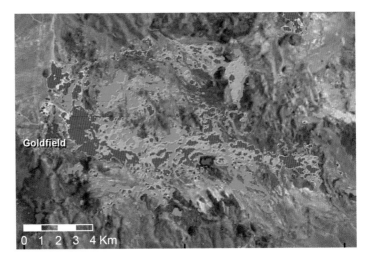

Goldfield

0 1 2 3 4 Km

Although silver can occur in many different types and ages of rock, in the United States it is more common in younger volcanic rocks (younger as in tens to hundreds of millions of years old). Exploration means zeroing in on a target, starting from a big picture of the general geological terrain and then progressing through ever-tightening rings of focus. The task is not so much to look for the elephant, but for signs of disturbance that might indicate the presence of the herd behind the trees – say, a flattening of grass, footprints at the waterhole or chewed bark. In other words, anomalies. Rather than looking for silver, geologists are looking for those subtle changes that might indicate the action of silver-containing fluids. For instance, when these fluids seep into limestone, which often surrounds silver or lead ores, the limestone is transformed into dolomite – a lighter and more coarsely textured rock.[9]

Some changes can be observed on the ground; others are detected from space. Volcanic rocks contain quartz and feldspar that can be readily broken down into clay by hydrothermal fluids. So the presence of certain types of clay minerals in bodies of volcanic rock may be a 'geophysical signature' for precious metals. To identify these altered rocks, geologists might use tools such as satellite imagery that can capture wavelengths of light invisible to the human eye. Different features of the earth's

surface show up at different wavelengths, and this information can be portrayed in psychedelically coloured maps that chart the earth's surface in literally a different light. For example, clay-altered volcanic rock at the historic gold- and silver-mining region of Goldfield and Tonopah in Nevada is pictured in stark red and green.[10] There, in colours that would have pleased a Fauvist, is the marrying of a millions-of-years-old hydrothermal plumbing system with space-age technology.

Coeur Rochester open-pit mine, Nevada.

Mining and refining

Exploration is only an infant step towards the creation of a successful silver-mining operation. Many political, economic and environmental factors have to be resolved before extraction begins. Silver mines are either open-cast or underground, depending on the location of the ore. Today's open-cast mines are radically transformed landscapes, resembling vast, denuded Southeast Asian rice paddies. Because historic mines often become tourist attractions, many of us are more familiar with underground mines, entered through vertical shafts, or horizontal adits – tunnels that burrow into hillsides. The earliest silver mines, which scraped away surface deposits, were pits or trenches. But shaft sinking is also an ancient technology. The Egyptians, mining gold in the Nubian Desert, sank circular shafts with footholds in the walls, sometimes to a depth of 30 m (100 ft).[11] Shafts during the Roman period could be as deep as 200 m (650 ft). The Industrial Revolution, though, advanced the depths to which shafts could sink, with the invention of steam-powered water pumps and hoists to lift ore to the surface. Today's underground behemoths plummet much further. A shaft at the Lucky Friday mine in Idaho's Silver Valley is projected to reach a depth of close to 2.6 km (1.6 miles).[12]

The ores produced from these mines require an extensive process of refinement before they yield anything like the precious metals we recognize. Silver is often found in combination with

Headframe, which houses the hoisting mechanism for transporting miners, equipment and ore, and ore bins, which store the mined ore before processing. Butte Mineyards, Silver Bow County, Montana.

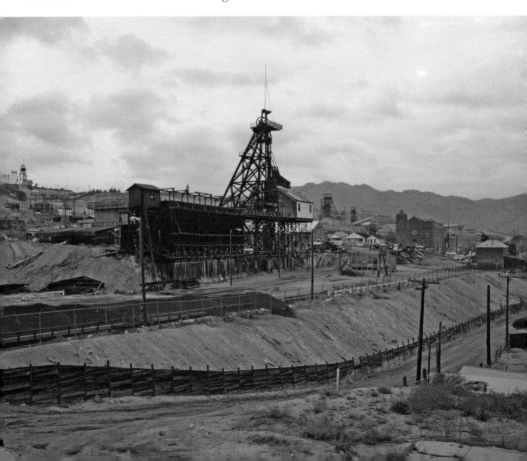

The processing facilities at Coeur Rochester in Nevada.

lead and zinc, and these sulphide ores can be treated by the flotation process, often at an on-site mill. The first step in recovering the precious metal is to grind the chunks of rock down to the consistency of flour. The fine material is mixed with water to form a slurry, and this is then treated with chemicals in flotation tanks to separate the minerals. Dry bars of silver-lead concentrate are then usually shipped to off-site smelting plants where the silver is further refined.

Another method of separating ore is through a heap leaching process, which involves an enormous footprint on the land. The leach pads, which might cover hundreds of acres, are essentially plastic liners on a clay base. Layers of ore are dumped on these liners, and are treated with a cyanide solution that dissolves out the silver and gold into a pregnant solution – a mixture that is then collected and processed to extract the precious metals.

Purity

Processing ore, refining and purifying are lengthy operations, and may involve plants in different countries. Given this complexity, how can we be confident in the purity of silver we might buy as an investment, or for ornament? It is possible to buy a 100-g (3.5-oz) bullion bar of 999 (99.9 per cent pure) silver, imprinted with a serial number for security, and whose weight is guaranteed by an Act of Parliament from the British Royal Mint. Throughout the world, authentication has often been ceded to the highest authorities. From the time that silver was understood as a store of wealth, ingenious measures have been employed to guarantee its purity. The earliest-known Western coinage – from Lydia, now in Turkey – was made from a mixture of gold and silver called electrum. We know from surviving coins dating from the seventh century BCE that the Lydians produced coins of consistent weight and purity in several denominations, with die-struck designs on the front and punch marks on the reverse. The image of the lion on these coins has led to the theory that they may have originated in a royal mint, a further guarantee of purity.

We can assume that over time silver-working cultures devised their own methods of quality assurance. For instance, the Byzantine emperor Anastasius established a system of stamping silver as an official guarantee of its quality, and to aid in the calculation of tax due. The system disappeared, probably when the administrative might of the empire declined.

The hallmarking that we are most familiar with today is a Western medieval convention that has proved admirably adaptable, and for historians concentrates a world of information. Hallmarks are like a sophisticated global positioning system for silver identification, with the added bonus of a temporal coordinate. The widely available 'sterling silver' (92.5 per cent silver, 7.5 per cent copper) is an old European standard in use from the twelfth and thirteenth centuries.[13] In Britain, the history of hallmarking reaches back to the thirteenth century, when a small group of London goldsmiths was officially tasked with ensuring standards of purity for gold and silver. Hallmarking certainly did not originate in Britain – older schemes in France used town marks and makers' marks – but the beauty of the British system was a simplicity and standardization that survives into the present. The leopard's head mark, still in use today to indicate London-assayed silver, gold and platinum, was adopted as early as 1300. In 1327 Edward III granted the Goldsmiths' Guild (later the Worshipful Company of Goldsmiths) a royal charter confirming its authority over assaying (establishing purity) and marking. Later statutes required makers to add their own unique marks, and dates were indicated by a letter of the alphabet, whose style changed with each 26-year cycle.[14] Assaying was to remain the privilege of the goldsmiths, a symbiosis neatly conveyed by the very word 'hallmark', referring to the Goldsmiths' Hall in London, where craftsmen brought their work to be assayed and marked. Since 1999 the British silver hallmark has been made up of three compulsory components: a symbol for the assay office (leopard's head for London, anchor for Birmingham, rose for Sheffield and the three-towered castle for Edinburgh); the purity mark: 800, 925 (sterling), 958 (Britannia) or 999; and the sponsor or maker's mark.

A very old means of assaying involved rubbing the article on a black touchstone and comparing the colour of the mark against samples of known purity. More accurate analysis could be achieved through cupellation – another ancient technique – which involved heating small samples in a furnace, extracting the silver and comparing its weight against the original sample. A recent non-destructive method uses an X-ray fluorescence spectrometer to determine the exact quantity of each element present. Once the purity has been established, the item can be hand-marked in

Marks on a skewer, by Hester Bateman, 1789–90, silver.

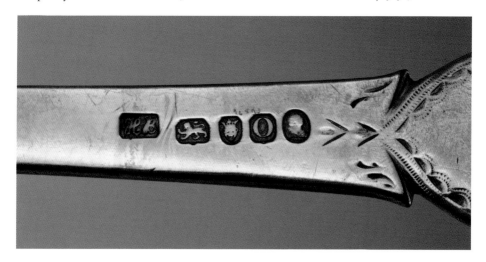

the traditional way with a steel punch and hammer, or laser-marked, which etches the motif onto the surface.

Every country has its own conventions, but a recurring theme is the trust placed in goldsmiths to ensure the purity of a country's coinage, and the integrity of consumer goods. In early Renaissance Italy, goldsmiths functioned not solely as artisans but as civil servants, overseeing the operations of mints and sometimes monitoring the circulation of coin.[15] A goldsmith's mark was a guarantee of the purity of the object, as well as of the integrity of the goldsmith.

Hallmarks are narratives that take place against political upheaval, the refashioning of borders, civic machinations and the waxing and waning of personal fortunes. These small, sometimes barely legible marks are a reminder of the much larger truth that our world is constantly in flux. Tens of thousands of hallmarks have been identified to date, with new ones added daily to the online databases and directories published throughout the world. On a piece of Austrian silver, for instance, the head of Diana, crowned with a crescent moon, tells us that this piece was created during the brief flowering of the powerful and populous Austro-Hungarian Empire. In 1867 the monarchy, with Habsburgian efficiency, imposed a uniform system of hallmarking for the entire multinational realm. Letters indicated the city of assay – A for Vienna, for example. By the end of the First World War, Austria-Hungary had collapsed. All those city letters – L for Ljubljana, M for Trieste, P for Pest, V for Zagreb – were suddenly the pieces of a shattered empire. In 1922 a much-reduced Republic of Austria introduced new marks with the slightly cartoonish profile of the hoopoe bird and toucan, instead of the regal Diana.

There is a Russian hallmark that existed for barely nine years. In an effort to overhaul a more cumbersome system of hallmarking, Tsar Nicholas II introduced the mark of a woman in profile, facing left, and wearing the traditional *kokoshnik* headdress. It lasted from 1899 to 1908, when a right-facing *kokoshnik* mark was implemented. Much could be made of that backward glance – this handful of years marked the publication of

Russian/Polish serving fork, made in Warsaw under Russian rule, showing a *kokoshnik* mark (top), silver.

Chekhov's *The Cherry Orchard*, with its poignant humour over the decline of the landowning class, and the Revolution of 1905: the beginning of the end. After 1927 the Soviet mark was of a worker in profile with a hammer, facing right into the future.

2 Silver Landscapes

Silver is found from Norway to New Zealand, from Alaska to Argentina – the latter being a country whose very name refers to the abundant silver the first European explorers hoped to find there. It occurs in vastly different landscapes and in underground terrains. These landscapes might be high in the snow-dusted Andean mountains; in damp, Scottish woodland; in the bamboo-crested hills of a Japanese island; or in the arid Australian outback.

In all these places, silver has been extracted through human ingenuity. The natural processes that bring us silver, occurring over hundreds of millions of years, have made their own impact on the terrain; human invention can reshape it again entirely. Silver landscapes are technological landscapes. In addition to exploiting existing technologies, silver mining has spurred new methods of extracting and processing ore and of refining the precious metal, all of which change the land.

While these landscapes are physical, they are also interior and sometimes supernatural. Silver is mined by men and women with aspirations and hopes. It has also been mined by slaves, and children forced underground by economic necessity. Whatever their circumstances, miners have developed beliefs and cultural practices to sustain them through their often difficult and dangerous labour.

Lastly, in harnessing technologies, reshaping terrain, consuming energy and introducing often toxic chemicals such as mercury and cyanide, silver landscapes are despoiled landscapes

whose impact on soil and water can extend for many kilometres (and many generations). Mining is a scarring process and whatever the success of reclamation efforts after operations have ceased, the environment is inevitably altered, and can never return to what it was before. Like everything else on our planet, silver is 'nuclear waste' from ancient stars that died long before the creation of earth.[1] Lest they lay waste to our environment, silver landscapes demand diligent restitution.

Silver mines at Aspen, Colorado, photochrom print, 1898–1905. The landscape was scarred by 19th-century mining schemes.

Technological landscapes

The earliest silver landscapes were lead landscapes. In fact, the earliest mining activities were not for metals at all, but for flint and obsidian with which to make tools. Metal mining probably

evolved as an extension of this, from experiments with rocks and minerals.[2] In the ancient world, most silver was derived from galena (lead sulphide), earth's most common lead mineral. It often contains enough silver to make it a viable source of the precious metal. Galena does not look like silver, but it looks enough like *something* to have attracted curiosity. It has a dull lustre, pronounced crystals (it can look like a wrenched-apart Rubik's Cube) and is surprisingly heavy. Lead was mined in Anatolia (a region that now makes up most of modern-day Turkey) from the seventh millennium BCE. One source was the mineral-rich Taurus Mountains, a range on the southern border of the Anatolian Plateau. Lead can be quite easily extracted from galena by smelting – that is, roasting the ore over a fire of charcoal or dry wood. Lead beads dating to the seventh millennium BCE have been discovered in Anatolia, and lead ornaments were sometimes used as luxury items. The ability to smelt lead, however, probably stimulated the extraction of silver, though this would have to wait at least a couple of millennia.

Silver objects containing trace amounts of lead have been found in Anatolia, Mesopotamia and Palestine, dating from the fourth millennium BCE. During the smelting of galena, any silver runs off with the lead. The next step is to separate these metals via cupellation. Using this process, the molten mixture is placed in a crucible and air is blown over at high temperature. The lead and any other impurities oxidize, and the silver is left behind. This early refining technique continued to be used for millennia, and, like smelting, was often practised in workshops close to the sites of ancient mines.

The Cyclades Islands in the Aegean produced artistically accomplished silver objects from the third millennium BCE. Today these Greek islands are best known for their iconic white villages and beach-fringed turquoise waters, but starting from the Early Bronze Age they had several small galena mines. The Greek historian Herodotus wrote that in the middle of the first millennium BCE, the island of Siphnos had the richest mine, and its wealth funded a marble treasury at the sacred site of Delphi to store religious offerings.[3] The few surviving delicate bracelets

and finely incised bowls from these islands reflect the Cyclades' rich culture and early advances in metallurgy.

The mine at Siphnos might have been rich, but the confines of geography and technology meant that it was small. Some ancient mines, however, such as the Roman mines at Rio Tinto in Andalucía, Spain, were enormous. Today the name Rio Tinto conjures the behemoth British–Australian global mining company that bought the Spanish property in the nineteenth century and turned it into the world's leading producer of copper. But Rio Tinto's riches were far more diverse. Sometimes clues to mineral wealth are hidden in landscapes and revealed through a

Kelley Mine, Butte, Silver Bow County, Montana, USA. Mining turned 'The Richest Hill on Earth' into a contaminated Superfund site, subject to costly clean-up. In recent years there have been efforts to integrate preservation and tourism into the historic mining environment.

32

Bowl, early Cycladic, *c.* 3200–2200 BCE, silver.

mixture of knowledge, intuition, patience and luck. At other times they lie on the surface. To those in search of minerals, the landscape of the Rio Tinto must have been gloriously inviting. Rocky hillsides are streaked gunmetal grey, blue and crimson, and the river is tinted rusty-red from iron. It attracted the attention of Copper Age inhabitants who extracted copper-rich ores. More substantial mines have been dated to the seventh century BCE, but mining on a colossal scale began after 206 BCE, when the Romans occupied the region after defeating its Carthaginian rulers. They turned Rio Tinto into the largest mine of antiquity.[4]

Over the following centuries, Roman engineers created a system of shafts sunk to 135 metres (440 ft) below the surface. These required ventilation and drainage systems. Remnants of pumping systems that used treadmill-operated wheels have been excavated from the site. With Roman bureaucratic efficiency, the mines were run by professional administrators with strata of labourers; some were skilled professional miners, but many were slaves. Some of this workforce lived in comfort. Rio Tinto was well provisioned with the material luxuries of a large

Roman town; there were bathhouses, potteries and cemeteries with roomy graves. Slaves, on the other hand, were chained at the neck and at the end of their short lives their remains were tossed on the slag heap.

Technological progress is enhanced by clear communication. The person who may have done most to initiate sharing of mining knowledge was neither an engineer nor mine administrator, but a doctor. When Georgius Agricola arrived in Joachimsthal to serve as a physician in 1527, the Bohemian mountain town (now

Rio Tinto mining landscape in Andalucía, Spain. By-products of millennia of mining have contributed to the river's acidic red waters.

Jáchymov in the Czech Republic) had been a richly productive centre of silver mining for a decade. The population was burgeoning, investment from European banks was flowing, and six hundred to eight hundred mining permits were issued each year.[5] The Joachimsthal mint began producing coins in 1519/20, most famously a high-value silver coin named the Joachimsthaler, shortened to thaler. It was imitated throughout much of Europe and in English called the dollar.

Agricola had studied classics at Leipzig University and had then taught Latin and Greek for several years before studying medicine. It was as a humanist scholar that he approached the prodigious task of cataloguing and communicating current knowledge of mining and metallurgy. His groundbreaking text *De re metallica*, published in 1556 shortly after his death, was the product of a lifetime of observation and first-hand experience of mines, contact with engineers and miners, and familiarity with contemporary research. *De re metallica* was published in Latin, the international language of scholarship, and illustrated with 292 woodcuts, the better to communicate technological achievement as broadly and clearly as possible.

A true Renaissance mind, Agricola debunked folk belief in favour of scientific observation, rejecting, for instance, the popular use of hazel twigs as divining rods in locating silver. Rather than trusting in 'enchanted twigs', a serious prospector should, according to Agricola, study the landscape for himself and observe changes after storms, erosion or fire, or in vegetation – not so very different from the methodical observation a geologist might undertake in the field today. He also believed that metals were deposited by fluids in rock fissures – a clear contribution to the modern theory of hydrothermal deposits.

It is from the woodcuts, prepared from the drawings of a Joachimsthal artist, that we get the liveliest idea of the impact of large-scale Renaissance mining on the landscape. Many are cross sections, depicting subterranean galleries and shafts, as well as the hive of activity on the surface. Agricola gives us several alternative technologies to address the age-old problem of pumping water from underground mines. Low-tech solutions included windlasses (crank-operated winches) to haul up buckets of water by hand. There were treadmills powered by horses. At the more sophisticated end, there were wheel-operated triple-suction pumps. As a physician, Agricola was doubtless concerned with the health effects of poor air that could suffocate or poison a

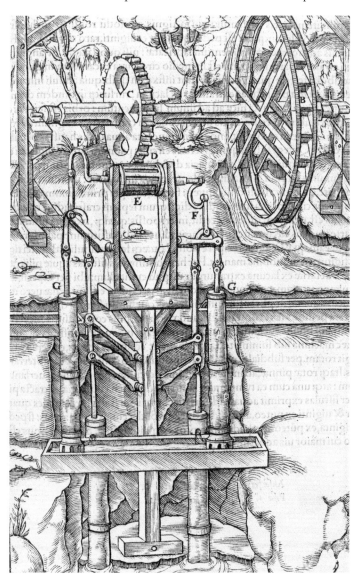

A mechanical water-powered pump, illustrated in Agricola, *De re metallica* (1556).

miner. He proposed wooden fan structures to blow wind down shafts, and systems of large leather bellows.

The grinding of ore, smelting and refining were conducted close to the mines, and required mills, furnaces (sometimes sophisticated multi-chamber structures) and systems of treatment tanks. On this industrial scale, a mining landscape was an above-ground terrain of chimneys, fire, smoke and chemical fumes. The living trees pictured in so many of the woodcuts were quickly anachronisms. Wood was required for mining equipment and tools, as well as for fuel for processing ore; the hills of Joachimsthal were quickly denuded, and by the 1540s the forests reaching up to the border of Saxony were cleared.

Deforestation has been a long-term consequence of mining. At times its effects have ranged across state and country borders; sometimes forests have been felled for mines on different continents. During the second half of the nineteenth century the legendary Comstock Lode in present-day Nevada poured silver into the economy of the American West, and stripped its forests of trees. 'The Comstock Lode may truthfully be said to be the tomb of the forests of the Sierras', wrote Dan De Quille, a Comstock reporter who drew public attention to the environmental degradation that accompanied the bonanza.[6] Twenty-four million metres (79 million ft) of timber were devoured on the Lode each year, a catastrophe that might, De Quille mused, mislead geologists of the future into believing the resultant coal fields were formed by massive beds of driftwood at the bottom of a lake. He painted a picture of a deranged landscape where giant flumes connected timber mills high on hillsides with valleys 3 km (2 miles) below – a demented reality mimicked in tame miniature by the log flume rides of today's theme parks.

Destructive as this was, much of this timber saved lives. Cave-ins presented a critical mining safety problem, exacerbated by the friable, shifting rock of the Comstock. A solution that revolutionized safety standards was devised by a young European engineer, Philip Deidesheimer, who, incidentally, had attended the prestigious Freiberg University of Mining – a

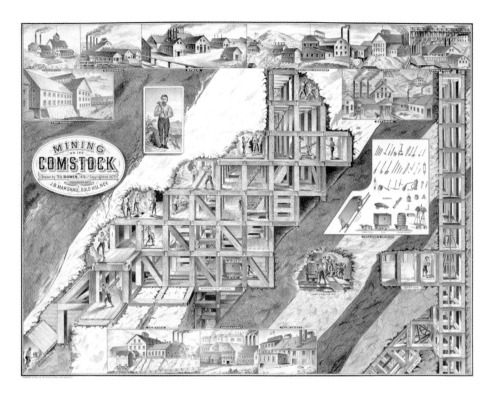

historic institution whose early curriculum was much influenced by the work of Agricola. Inspired by the structure of honeycombs, Deidesheimer invented a system of stacked timber cubes that came to be known as square set timbering. These modular timber cages allowed miners to safely excavate the generously wide veins characteristic of the bountiful Comstock Lode.

Another reporter on the Lode, and a colleague of Dan De Quille, was the young Mark Twain. A brief stint as a miner acquainted him with the vast underground world of timbering that resembled 'the clean picked ribs and bones of some colossal skeleton'.[7] The miles of timbering turned the mine into an underworld city of boulevards and soaring spaces taller than any city cathedral. The Spanish proverb he quoted – 'It requires a gold mine to run a silver one' – is an apt comment on what was soon to follow on the other side of the world, in a place as close to the middle of nowhere as any mine has been located.

T. L. Dawes, 'Mining on the Comstock', 1876, lithograph.

On the dusty, hot plain of the New South Wales outback, the Barrier Ranges are a crease on a smooth sheet. Despite the remote location, early prospectors found them promising targets; they glittered with white quartz, and rocks were pocked a tantalizing green. Reports of copper and silver began to trickle in during the 1870s and '80s. One long, low hill with a jagged profile was named Broken Hill. Its crest, over a kilometre (half a mile) long, was capped by an 'iron hat' of manganese and iron oxides. For those hoping for silver, it seemed a disappointment, but optimistic prospectors continued to explore. After several years, some promising silver deposits were discovered in clay, and the rush began in earnest. Thousands poured into the camp, and shares surged, but faced with the sudden influx, Broken Hill's aridity and isolation almost proved its undoing. Water was scarce, sanitation scarcer and typhoid a scourge. The shafts and tunnels were deathtraps, reinforced with native gum trees that rapidly decayed. Sensibly, the mines' owners looked to Nevada for guidance, and shipped over (for handsome fees) engineers and miners from Comstock. Square set timbering was constructed inside Broken Hill's tunnels to protect miners from the crumbling, unstable rock, but the lumber came not from Australia but from the United States.[8] Shiploads of Oregon pine departed from West Coast ports and crossed the Pacific and the equator to line the bowels of outback mines. Although Broken Hill was one of the world's richest silver–lead–zinc deposits, it took the metaphoric gold mine to run it.

Supernatural landscape

Working conditions in underground silver mines are dark, noisy, hot, humid and dangerous. At 1,000 metres (3,280 ft) below the surface, temperatures can easily climb to 50°c (120°F). Hazards range from heat exhaustion to cave-ins, arsenic and mercury exposure, carbon monoxide poisoning, fire, equipment accidents, scalding by hot water and lung damage from inhaling the silica dust that hangs in clouds after drilling. Little wonder, then, that mining communities developed a rich store of beliefs and

Ancient remains of Iwami Ginzan silver mine, Honshu, Japan. The mine produced silver from the 16th to the 19th century.

Cages for transporting miners underground, suspended beneath the Badger State Mine headframe, Butte Mineyards, Montana.

practices to help them locate precious resources and keep miners safe underground. Throughout many cultures runs the belief that the geological realm is a sacred space quite distinct from the world above ground.[9] In this realm, underground deities reveal the hidden veins of silver and gold. At Finistère in Brittany, for example, a fairy was said to have led humans to the rich silver-bearing lead ore which was mined by Celts and Romans. These underground spirits were exigent, though, and were placated only through constant attentiveness and the offering of gifts. To facilitate the appeasement, shrines were commonly built near mines; shrines to the Egyptian goddess of miners, Hathor, have been found close to mines at Timna and in the Sinai Desert.

Japan's richest silver mines at Iwami Ginzan, on the main island of Honshu, were also reputedly discovered through divine intervention. In the early sixteenth century the wealthy merchant Kamiya Jutei was sailing on the Sea of Japan when he was diverted by a holy light which beckoned him into the

mountains.[10] Iwami Ginzan became one of the world's most bountiful silver mines; at the height of its productivity in the early seventeenth century, it produced one-third of the world's silver. The site comprised about 150 villages, and the production of silver was closely integrated with daily life and religious practice. In addition to Buddhist temples, there were numerous shrines near pitheads and strewn along the transportation routes to shipping ports. Some temples housed the guardian deities of the mines; other shrines were noticeably more specific about their prophylactic benefits: there were, for instance, shrines for fire protection, for a healthy supply of drinking water and for maritime safety at the ports where the silver was sent on its route to Korea and eventually China. These temples and shrines played an integral role in the mining community and were (and still are) the sites of large festivals.

The silver mines whose supernatural landscapes are most daunting are probably those at Potosí in the Bolivian Andes.

Potosí, Bolivia, 2007.

Exploited by the Spanish conquistadores from the middle of the sixteenth century, Potosí was likely the world's largest city in the early seventeenth century, bankrolling the emergence of Spain as a world power. At a lung-punishing 4,090 m (13,420 ft) above sea level, Potosí is among the world's highest cities. It sits at the foot of Cerro Rico – the Rich Hill. Geology explains how the conical mountain originated as a volcano with a rock core that had been permeated by hydrothermal solutions rich in silver and zinc. The dome of this rock was exposed during volcanic eruptions, revealing the rich veins of silver that lace the mountain.[11]

But to the Quechua Indians who inhabited these mountains, this bizarre cone was otherworldly. In Andean cosmology, mountains were the sacred places that birthed the founders of the human race. Potosí may be named after the Quechua word for thunder, or after the cataclysmic storms that regularly surge through the mountains, but it also has an echo in a local legend telling of a thunderous voice that warned men away from the hill's riches. The violation of the mountain, then, was a cultural trespass that demanded serious propitiation.

That propitiation was found in El Tío. In every mine on Cerro Rico, he sits in a cavern lit by lanterns whose illumination is sinister rather than reassuring. Although El Tío is the Spanish term for 'uncle', this grotesque, disfigured statue inspires fear rather than affection. Horns twist above eyes that gape into the maw of 'the mine that eats men'. At first glance, his mouth appears crammed with fangs; but these are cigarettes stuffed into a hole below an ill-kempt moustache. His large erect penis is like a third fang jutting from a lap heaped with gifts of brandy and rum to slake his insatiable thirst for tribute, and coca leaves to pacify his bellicosity. The sexual reference is a gesture, too, to Pachamama, the Andean earth mother who birthed these mountains.

Above ground, the landscape of Potosí is bleak and wind-scoured. Ironically now one of the world's poorest cities, it is a panorama of low, dun-coloured houses interrupted by the many steeples and domes of its colonial-era churches and monasteries. Potosí is a city infused by its Roman Catholic past. The mountain's silver fitted out these churches, and although some of it has migrated to the international marketplace, some remains in silver-coated icons of the Virgin and Child, shaped like the triangle of Cerro Rico.

The miners, though, inhabit a more Manichean universe. While there are crucifixes outside the mine entrances, and miners cross themselves and pray as they leave their homes, once they are underground, they inhabit the dominion of the devilish El Tío, who owns the minerals:

> Outside we believe in God who is our only Saviour. But when we enter the mines things change. We are entering the world of Satan, the Devil. We ask him for favours, sometimes on our knees, lighting candles for him, so our belief is split into two worlds.[12]

El Tío, Potosí silver mines, Bolivia.

In this worldview, 'blessing' is required of both a Christian God and El Tío. In the 2005 documentary *The Devil's Miner*, fourteen-year-old Basilio Vargas, one of the estimated eight hundred children who labour in the mines, says, 'Only if the devil is generous will he give us a good vein of silver and let us get out alive.' The days of high-grade ore are long gone, and sometimes the gifts of alcohol and coca (which the miners chew to combat hunger and fatigue) are not enough. At the start of each August, a llama is sacrificed as an additional payment to El Tío.[13] On these days, the dichotomy of the miners' beliefs is at its most startling. After attending church, the miners return to decorate El Tío and slaughter a llama, whose blood is thrown around the mine entrance in an echo of Passover, splashing the crucifix. 'Without these sacrifices he would kill you. He would take the sacrifice from your own flesh.'

A prominent feature of above-ground Potosí is the packed cemetery; the mines have killed an estimated eight million people. It's a necropolis of concrete monuments bristling with simple crosses. The most sympathetic commentary on the interior landscape of Potosí, though, comes from the contemporary Bolivian writer Victor Montoya. A native of Potosí, the unbearable conditions there led him to become a political activist and advocate for reform. In an essay addressed to El Tío, he acknowledged that the miners have been locked into lives of fear and superstition, but also the narcotic draw of the 'uncle': 'I have the terrible sensation that you are pursuing me as if you were my shadow. Sometimes you are closer to me than Faust's Mephistopheles.'[14]

Landscapes of hope and disappointment

Some silver mines are prodigious; at others, the output is paltry. Throughout the history of metal mining, promising discoveries have inspired rushes, the building of boom towns, the creation of wealth and the transformation of economies. But those enriched by silver mining have far been outnumbered by those disappointed or even ruined by their investments. The history of silver rushes may glitter with exhilarating stories of profit, but it is also littered with tales of tragedy and loss. Prospectors, miners and investors all bore the brunt of failed expectations, but the landscape also bears the marks of these disappointments. Across the world, from the Australian outback to the Colorado mountains, there are abandoned 'Silvertons' bearing hope in their names, and ruin in their remains.

The abundance of silver that flowed from the American West is legendary; the abundance of its abandoned silver mines, while gladdening the spelunker, is an acknowledged environmental hazard. The first great American silver rush began in the 1850s. Prospectors from eastern states and from abroad began converging on California following the discovery of gold at Sutter's Mill in 1848. For silver mining, the bonanza arrived in 1859 with the opening of the spectacularly bountiful Comstock Lode in present-day Nevada. The flow of silver from Comstock and other Western mines made San Francisco and many other parts of the United States rich, and precipitated a frenzy of prospecting and stock market speculation in the American West. Comstock overshadows the story of the myriad other small mining operations in Nevada and California, but the plot tends to repeat like the echo of a dynamite blast in a tunnel. As mining companies became established and mountain slopes crawled with mule teams hauling ever-richer loads of ore, expectations grew that anyone with a modicum of courage and ready cash could profit.

But the geology that primed conditions for the creation of mines could also prove their undoing. If silver gathered in sediments near the earth's surface, weathering and erosion could give the impression of a far richer hoard of silver than was the case. A mine could quickly founder after this 'low-lying fruit' was harvested. Further, in earthquake-prone California, numerous fractures and faults made it extremely difficult for miners to follow the veins of silver. A mine could become unworkable despite a large initial investment in its creation.

California now has nearly 40,000 abandoned mines. Most of these were metal mines, and many were deserted shortly after

their creation, when insufficient minerals were found. Others
closed when silver prices plunged later in the century. California
may be the Golden State, but it is also a landscape of disappointment, whose arid mountains are riddled with forgotten
shafts and galleries. Tucked into the canyons are small communities that owe their beginnings, though not their ongoing
sustainability, to the silver rush.

One such community is Silverado, a small town dating from
the 1870s, in a canyon nestled in the foothills of the Santa Ana
Mountains, 80 km (50 miles) southeast of Los Angeles. The
name itself implies so much promise and misjudgement: a
corrupted riff on the Spanish El Dorado, the Golden One, the
gilded king, or the city of gold sought so arduously – and fruit-
lessly – by the Spanish conquistadores in South America.
Silverado wasn't always the name of the small canyon. Before the
development of the mines, it was known in Spanish as Cañón
de la Madera (Timber Canyon). And before that, by words in
the Tongva and Acjachemen languages for villages and places of
significance to the soon-to-be-displaced native peoples. But the
1870s was an era of great expectations, and Silverado it became.
It wasn't to enjoy a unique name, either. North of San Francisco,
on the slopes of Mount St Helena, another Silverado had
recently been established and quickly abandoned when it failed
to prosper. Financially embarrassed, Robert Louis Stevenson
and his wife Fanny spent their honeymoon there in 1880 in a
derelict miners' bunkhouse. *The Silverado Squatters*, his account
from 1883 of their experiences, earned the area enough renown
later to be designated Robert Louis Stevenson State Park.

Southern California's Silverado is now a small, rural community. Wooden homes, many dating from the 1930s and 1940s,
when the area was a modest resort, are slotted into the spaces
between ravines. The slopes are fragrant with sage and the whiff
of counterculture. Parked Harleys outside the Silverado Café
lend a cast of classic Western Americana. In terms of the mines'
output, Silverado was a small chapter in the story of America's
silver rush. But here in microcosm was the whole cycle of boom
and bust, discovery and depletion, hope and disappointment.

Silverado's story encapsulates all the frenzy and euphoria that silver inspires. It gives us a snapshot rather than a formal portrait of the transformative power of silver on our natural and cultural landscapes. But snapshots often shed much light on the impact of momentous change on ordinary lives.

Bedford Canyon Formation, Santa Ana Mountains, California, 2012.

According to local lore, a piece of silver-bearing blue-white quartz serendipitously picked up by a couple of hunters precipitated the rush on Silverado.[15] The history of precious metal mining is full of such tales of lucky accident. The founding of Silverado's most successful lode is a story of even more outrageous good fortune, vaguely reminiscent of the sort of Western movie shot by Hollywood when it later discovered Silverado as a location. John Dunlap, a U.S. marshal, arrived in Silverado on the trail of a Mexican fugitive staked out in the mountains. Pursuing his prey down a remote branch of the canyon, he instead happened upon silver; the outlaw remained at large.

The emphasis on happy accident is almost always misleading. Then, as now, prospectors hunted 'in elephant country', scouring the sites of prior mining activity. For the previous several decades, Spanish and Mexican settlers had been mining the southern Californian mountains; a major reason behind the Spanish colonization of the state was for its minerals. Nineteenth-century prospectors had a hunch where to look, and could recognize common minerals such as galena. Although they didn't have the advantage of present geological knowledge, there was an understanding of the metals that were likely to occur together, such as silver and lead – and that good places to start looking were the meeting points between metamorphic rock (rock that has been transformed by great heat or pressure) and igneous rock, formed after magma cools and solidifies.[16] One such visual attention-grabber lay in the canyons surrounding Silverado. Here, a Jurassic-era, partly metamorphosed assemblage of sandstone, quartzite, slate and shale – strikingly visible in places like a layered millefeuille (or Napoleon pastry) – comes into contact with a younger formation of volcanic rock.

After the first promising rocks were assayed in 1877, and found to contain enough silver to justify the cost of extraction and refining, hundreds of hopeful miners and their families converged on Silverado Canyon to stake claims. At first, they lived in tents; before long, a makeshift town of bunkhouses, hotels, markets, blacksmiths, saloons and brothels was in full swing, the mountain slopes were busy with mule teams, and stagecoaches ran twice daily to Los Angeles.

While hopeful prospectors and miners brought with them their courage and endurance, even the most rudimentary mine needs some capital investment. The history of America's silver rush is intimately tied to the history of the American West's stock market. Needing to raise capital for exploration, technology and daily operations, the new mining companies throughout the West issued shares that were released to the general public. Some of these shares began to pay dividends, spurring an epidemic of speculation among everyone, it seemed, from bankers to clergymen, cooks and day labourers.[17] In a heady atmosphere of euphoria and poor regulation, abuse was rampant. Unscrupulous brokers – the founders of the San Francisco Stock and Exchange

Board were dubbed the Forty Thieves – preyed on the naivety and greed of unseasoned investors. The gullible, with little concept of financial risk, were persuaded into short sales and margin sales. One particularly burdensome ploy was the leveraging of assessments on shareholders' existing stock, rather like a recurring property tax. In many cases, brokers and mine management simply pocketed the cash. The unfortunate holder could either choose to continue to pay in the hopes of eventual return, or default – in which case he forfeited his stock. The suicide rate in San Francisco nearly doubled during this period. Even Henry Comstock, after whom the Comstock Lode was named, shot himself after losing all his wealth to speculation. In retrospect, statistics suggest that euphoria was a mass inoculation against common sense. In 1879, at the end of the boom created by the Comstock 'Big Bonanza' mine, $24,000,000,000 in shares had been released. This was ten times more than the worth of all the silver and gold mined in the American West up to this date, and equal to the wealth of the entire nation.[18]

Fraud also may have tainted the history of the northern California Silverado mine where Robert Louis Stevenson honeymooned. According to one account, the mine had been a 'majestic swindle' created as a front for the sale of worthless stock. By night, trains of packhorses laden with old cigar boxes were used to smuggle silver up the mountain and into the mine, where it was crushed with local rock and sent back down as the mine's bounty.[19]

In Southern California's Silverado, the most successful mine, the Blue Light, did attempt to issue a block of 50,000 shares.

Sample of ore sent by Silverado mine owner John Dunlap to Mendheim and Hofmann, San Francisco, c. 1879.

The first were bought by the mine owner and managers, but only 82 shares were ever sold.[20] Silverado enjoyed a short boom, but most miners gave up in discouragement – a pattern repeated throughout the West. The silver, gold and copper from Western mines indubitably boosted the economy and made their owners among the richest men in America. But nearly all the wealth was created by less than one-tenth of the mines.

Reclaimed landscapes

The Greenland Ice Cap is a time machine, allowing us to journey back into the distant past and learn about the air our ancestors have breathed over the last 9,000 years. While we might assume poor air quality is a consequence of the Industrial Revolution, core samples of ice tell a different story. Toxic air, particularly from lead, has been traced back over two millennia. All lead-containing ore has its own chemical signature, and written in the Greenland ice is the proof that during the Roman era, toxins in the air increased substantially. With the exquisite exactitude of forensic scientists, researchers have traced 70 per cent of the lead in the ice-core between 150 BCE and 50 CE to the Rio Tinto mines.[21] In addition to polluting the air, Roman mining left millions of tonnes of contaminated slag from smelting operations. The zinc, arsenic and cadmium seeped into the soil, rivers and groundwater.

Rio Tinto demonstrates that silver mines can leave a legacy of pollution that extends over hundreds of years and thousands of kilometres. Toxins, if they were recognized, were generally accepted as a by-product of economic progress – at least by those who lived long enough to enjoy that progress. Dissenting voices tended to be lonely ones. Although Dan De Quille was certainly a supporter of mining operations in Comstock in the 1870s, he cautioned of other side effects in addition to deforestation. He puzzled, for instance, over the 7 million tonnes of mercury that had entered Nevada over the previous ten years, but had seemingly disappeared. Mercury was used to extract silver from its ores in an operation called the Patio

Process, an alternative to smelting. Mercury wasn't used up in the process, so it must have gone *somewhere*, he reasoned. As the U.S. Environmental Protection Agency confirmed more than a hundred years later, that somewhere was the bed of the Carson River, still so contaminated with mercury, arsenic and zinc that it has been designated a Superfund site (a site operated by the U.S. Environmental Protection Agency to identify and clean up hazardous waste sites in America), which requires long-term clean-up.[22]

It is only in the very recent past that legislation has required mine operators to restore the land at the end of operations. In the United States, it was not until the 1970s that environmental legislation encompassing mining became more rigorous. A plan for reclamation and a bond to ensure available funds are now required before mining activities begin. However, this is not standard practice everywhere in the world, and even in developed countries bankruptcy can sometimes shift the burden of reclamation to the public purse.

Multinational mining corporations now foreground their reclamation efforts and readily communicate plans for the restoration of their sites. Two mine closures, in very different topographies, illustrate the steps that might be taken when a mine reaches the end of its productive life. The Golden Cross Mine on New Zealand's North Island was a gold and silver mine that operated from 1991 to 1998 – a typically short lifespan for today's precious metal mines. It was surrounded by lush countryside, rolling farmland and dense forest, and sat at the headwaters of the Waitekauri River; in such an idyllic setting, reclamation would have to address the aesthetic as well as the environmental impact. The Golden Cross was an open-cast as well as underground mine, and the most obvious scar on the landscape was a deep, terraced crater, a discordant ochre hole in a patchwork of green. The pit was capped and sown with vegetation, and the waste rock (5 million tonnes of ore) was placed into specially engineered disposal sites to prevent contaminants leaching out into the soil. Cyanide was removed from the tailings (the refuse and effluents created through ore-processing),

The Golden Cross Mine after rehabilitation, New Zealand.

and the surface water was allowed to discharge back into a stream now clean enough to support a trout fishery.[23]

In New Zealand, nature lent a helping hand in the form of 3 metres (10 ft) of rainfall annually. Within a few years, open gashes were transformed into luxuriant grazing pastures and wetlands. At El Indio in the Chilean Andes, at 4,000 metres (13,000 ft) above sea level, reclamation presented different challenges. Like the Golden Cross, El Indio was an open-cast and underground mine. But it lay in arid, rocky terrain that could never repair itself with vigorous vegetation regrowth. More dramatically, a river had been rerouted to accommodate the mine at the start of operations. Chile lacked the strict environmental legislation of New Zealand, but the mine's multinational operating company, Barrick, implemented a reclamation plan that provided a model for future efforts. The slopes were reinforced

and stabilized, contaminated rock was hauled away for treatment, and the river returned to its original course.[24]

Calico Ghost Town, San Bernardino County, California.

Restoring landscape is not the only option, though; sometimes reclamation means turning the site into something radically different. The site of the Homestake Mine in South Dakota, formerly a deep gold and silver mine, now houses the Sanford Lab, a state-of-the-art underground research facility. Shielded by a mile of rock that protects experiments from distorting cosmic rays, physicists enquire – quite aptly – into what happens in stars when they die.

Despoliation aside, abandoned silver mines can have a romantic lure. They can be both terrible and tantalizing, as Robert Louis Stevenson attested after his Silverado honeymoon in his *Silverado Squatters*. The 'world of wreck and rust' where an abandoned iron chute hovered over the canyon 'like a monstrous gargoyle' was also one that worked magic on the imagination. In regions with many abandoned mines, authorities can either invest considerable energy and expense in keeping people out, or they can invite them in. The mountains of California that experienced the gold and silver rush are now full of such 'ghost towns' that have been transformed into tourist sites. Typical attractions include not only a tour of a mine in a rattling ore car, but sarsaparilla parlours; mule team rides; ghost tours with an emphasis on gunslingers, charlatans and prostitutes; and, of course, brothels festooned with red petticoats and long johns.

While entertaining, such 'reconstructions' by the heritage industry can also be archaeologically valuable. Iwami Ginzan in Japan had a four-hundred-year-long history as a silver-mining region. Near the southern tip of the main island, just inland from the Sea of Japan, the archaeological remains of 150 villages, six hundred mine shafts and pits, refining areas, temples and cemeteries are scattered in an extensive landscape of forested mountains and deep river ravines. The rains come over the

Stone bridge at Iwami Ginzan silver mine, Honshu, Japan.

hillsides in veils and nourish the thick bamboo and pine that screens the adits and levels former pits. Stone bridges arc over mountain streams, and ancient trails wide enough to accommodate a horse laden with refined silver wind towards ports that look out to Korea and the rest of the world.

Iwami Ginzan was added to UNESCO's World Heritage List in 2007. A far more reclusive ruin than many on that list, it nevertheless drew nearly a million visitors in the year following its designation. Often those who promote silver landscapes for tourism are aided by superb topography and conjure further interest from dubious reconstructions and popular imagination. In the case of Iwami Ginzan, what survived was a 'relict landscape' preserving every step of silver mining from extraction to the grinding of ore and smelting. Remnants of buildings from the seventeenth to the nineteenth centuries survived in villages, revealing a society that included soldiers (the site was fortified), rich merchants, many priests, prosperous family enterprises and farmers. At the height of its productivity in the early seventeenth century, 10,000 people were employed in mining. Written, however obscurely, on these hillsides was the story of how numerous small, labour-intensive operations multiplied to create Japan's most productive silver-mining region, with a flourishing trade network that extended through East Asia. As UNESCO emphasized, the special geography underlying silver deposits also creates a platform for profoundly cultural landscapes.

3 Silver Transformed

Today's silver ring might have been a part of Louis xiv's silver throne at Versailles, which long before might have been a hoard of Roman silver coins, which in their turn might have been minted from silver dug out of a mine in Andalucía. Silversmiths reshape a material that has already undergone aeons of transformation. And, however beautiful or skilfully wrought, the objects they create are probably not the final stage in silver's journey, but instead just another step in an ongoing process. If we consider all the silver objects that have ever been made (from coins to jewellery, to tableware), or all the objects utilizing silver (from photographs to electronics), most of these things no longer exist. They turn into other objects. We know this from historical records detailing the melting of luxury objects in times of economic need. We know it from annual statistics on the amount of recycled silver re-entering the market. Or sometimes the objects themselves give us clues about their origins. For example, the percentage of silver in neck rings created by Miao silversmiths in Guizhou Province, China, is sometimes exactly the same as coinage historically used in that region.

Perhaps it is more accurate to think of a silver object as a hiatus in the life of a material full of potential to be changed into something else. While it is the nature of silver to be malleable and easily melted, it is human nature to desire novelty and to feel need. Both these natures underlie silver's transformations in the hands of the smith.

Shaping

Raising

The most basic, commonly used technique to form a silver object is to hammer the silver over a stake – a process called raising. The silversmith starts with an ingot (a cast block of silver) or, more commonly now, a sheet. As a soft metal, silver can be hammered cold to stretch it gently into the form of a bowl, or hollow vessel. Over time, smiths created ingenious tools to help form the vessel in intricate ways. With extensive work, the crystalline structure of the metal becomes stressed and less malleable. Heating it and allowing it to cool (annealing) reduces the stress and renders the metal soft and pliable again. Some of the earliest surviving silver artefacts were raised, and the technique has been used throughout the world. For instance, the Cycladic bowl from Chapter Two was made this way, as were a collection of silver bowls

Necklace set, Miao minority, Guizhou Province, China, silver.

from the first and second centuries BCE, made in the region that is now Iran. As soon as humans learned how to shape vessels, it seems they yearned to decorate them in sophisticated and whimsical ways.

Cutting and joining

Raising requires both skill and time, but by the late eighteenth century, a new technology provided an easier way to form elegant shapes. Silver sheets, rolled to uniform thickness by mechanical presses, became widely available. The smith could cut shapes from sheet silver and join them together using silver solder – a metal alloy with a lower melting point than silver, which when heated would fuse with each surface. Paul Revere, the celebrated American silversmith-cum-revolutionary, purchased a flatting mill in 1785 and used it to produce Neoclassical oval teapots that were much sought after following the Revolutionary War.

Bowl, Near Eastern (Parthian), 2nd century BCE, silver, gilding.

Casting

The ability to cast molten silver was closely related to the ability to smelt silver from ore. In the earliest type of casting, 'open moulding', a shape would be pressed into sand or clay, or carved into stone, and molten silver poured in, like pouring water into a tray to make ice cubes. The disadvantage of this method, of course, was that one side would be flat.

A far more sophisticated method of casting, in use since the fourth millennium BCE, was the lost-wax process. A full-size wax model was created, and then coated with plaster or clay. When heated, the wax melted out through specially fitted channels, and molten silver was poured in to replace it. Once the metal cooled, the mould could be broken away. This was the method common through antiquity, and was almost certainly used to create the jaunty Greek Triton who originally formed the handle of a silver wine jug in the Getty Museum. The Triton (half man, half sea creature) hovering at the tip of a fish tail, would have faced inwards over the rim of the jug. His head and arms are solid, though his torso is hollow and might have been created through a refinement of the lost-wax process called core casting. The fish tail, cunningly curved into a handle, is a hollow tube onto which a solid cast flipper is fixed.[1]

Casting was the perfect method for creating the naturalistic sculptures of game birds, shellfish, fruits, vegetables and flowers found on French Rococo silver tableware. Several of the royal goldsmiths of the eighteenth-century French court may also have been accomplished sculptors.[2] This would certainly have continued a tradition dating back to the

'Triton' wine jug handle, Greek, 100–50 BCE, silver.

Paul Revere, teapot, 1796.

Mielle Harvey,
Bees Enter Box
(small wearable
sculpture), 2011,
lost wax cast silver,
oil paint.

Jacques-Nicolas Roettiers, tureen with cast neoclassical decoration, 1775–6, silver.

Italian Renaissance, when goldsmiths trained as artists, studied classical sculpture and could switch fluidly between media.

Specific casting techniques could be used to express an artist's cultural identity. For instance, since the late 1800s Native American silversmiths in the southwest United States have made casts from tufa stone, a locally found volcanic rock. Intricate designs can be cut into tufa's soft and porous surface, and silver, at first obtained from melted American or Mexican silver coins, poured into the mould. The finished surface retains the grainy texture of the stone, locking in a unique place and culture.

Anthony Lovato, horse cuff, silver, tufa cast.

Experiments with new technologies in the late twentieth century have introduced new materials for casting, such as precious-metal clay, which consists of silver particles suspended in an organic binder. Soft and pliable, the clay can be pressed into silicone moulds. Once dry, the clay is removed and fired, burning off the binder to leave behind a solid silver form.

Die-stamping

Imprinting a shape and design onto a piece of silver was a technology used to create the earliest coins. The production of larger pieces, though, had to wait until the Industrial Revolution and improvements in the steel used to make dies and advances in energy for powering heavy machinery. From the mid-eighteenth century, the metalworking industries became increasingly industrialized. Rather than a network of workshops concentrating on tasks such as raising, engraving and polishing, there now existed factories created by entrepreneurs, where all these processes could be housed in one location. Powerful presses stamped out multiple components that could be soldered together to form objects as large as candlesticks. By the nineteenth century, these factories were producing a plethora of domestic silverware and jewellery much more swiftly and at a lower cost than had previously been possible in workshops.

John Carter,
candlestick, die-
stamped and soldered,
1776–7, silver.

Filigree and wire

Filigree resembles lace made from precious metal, and can be found on silverwork as diverse as Viking bracelets, medieval crosses, Indian ornaments and fittings for furniture, Victorian buttons and contemporary Southeast Asian jewellery. The technique consists of working silver wire into delicate scrolls or tracery and soldering them to a frame. Sometimes an object can be formed entirely of filigree, or it can be applied as decoration. The airy effects of filigree meant that it was particularly favoured for women's ornaments; it enlivened brooches in Norway, waist clips in the lands of the former Ottoman Empire and necklaces in the Indian state of Rajasthan.

The state of Odisha, on the east coast of India, has been an important centre for Indian filigree work for several centuries. Although filigree is now a popular tourist item, traditionally it served a religious function and can still be seen adorning many temples. A common traditional bridal gift was a set of filigree dishes for making offerings at religious festivals.

Filigree was chosen as an appropriate technique for religious expression and also favoured by Yemenite brides. Prior to the creation of the State of Israel in 1948, Yemen had a sizeable Jewish population, and silversmithing was an occupation practised almost exclusively by this group, not just for Jews, but for Muslims and Arabs.[3] Since pre-Islamic times, they had developed a highly sophisticated form of filigree for jewellery and ritual objects, handing down knowledge from father to son.[4] When Yemeni Jews emigrated to Israel they created Judaica and the virtuosity of the gossamer-fine filigree was itself a form of worship. However, much of the craft was lost as the creation of jewellery was secondary to the immediate needs of the new state, and now it is practised by only a few artists.

Ben-Zion David, silver multi-cup *Kiddush* cup.

The Miao minority in southwest China are another group to express identity through filigree. The technique was probably adopted during the Tang Dynasty (618–907), when cultural exchange between China and other parts of Asia reached new heights. Tang silversmiths produced magnificent filigree jewellery for aristocratic use, but minority groups such as the Miao,

Ben-Zion David working at his table.

Bracelets, Miao
minority, Guizhou
Province, China.

who often crossed wide geographic areas, used jewellery as a way
of celebrating and continuing their culture and traditions. Most
silverwork has been made by male artists in silversmithing villages, and is produced for both everyday wear and festivals. Some
of the most common motifs – butterflies and spirals – embody
the Miao creation myth in which a butterfly mates with a water
bubble, and produces the first human ancestors. Miao silverwork
sometimes incorporates granulation, a technique often found in
conjunction with filigree. Small balls of silver are attached to the
filigree surface, heightening the effect of intricacy.

Perhaps because of its ethereal appearance, filigree has often
been used in the service of religion. Sometimes, though, its
applications are purely secular. One of Catherine the Great's

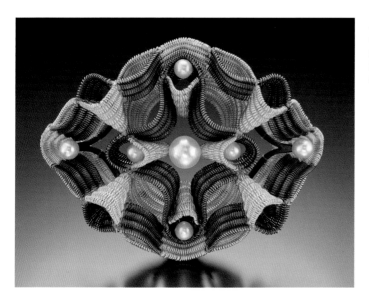

Anastasia Azure,
Coaxial Providence,
2009, sterling silver,
fine silver, fishing line
and pearls.

many appetites was for Chinese silver created in Canton and exported to the West. The unrivalled collection of Chinese filigree now owned by St Petersburg's Hermitage Museum includes items from Catherine's vast silver toilet set originally kept at the Winter Palace. Perhaps the most astonishing of the items collected by this consummate connoisseur are two silver crabs made to contain lipstick and rouge. They sit on filigree trays resembling delicate fronds of ribbed seaweed edged in blossoms.

Contemporary jewellers are experimenting with textile techniques that take filigree one step further. As filigree is worked with wire sometimes fine enough to be thread-like, the logical next step has been to manipulate fine, malleable silver wire to create textile forms such as crochet, weaving or knitting.

Embellishing

Repoussé and chasing

Repoussé is the art of raising a design from the reverse with hammers and punches, while chasing is worked from the front. Often used in combination, they create a multidimensional surface that can resemble sculpture. Repoussé and chasing were

The Gundestrup
Cauldron, possibly
Romania or Bulgaria,
100 BCE–1 CE, repoussé
silver with gilding.

known from antiquity, and worked in both silver and gold. Roman silversmiths were adept at these techniques and used them expertly to craft the Bacchic scenes that enliven so much Roman tableware: hovering cupids, frolicking satyrs festooned with garlands laden with pomegranates and grapes, birds whose beaks jut into our space, vines that flutter towards us – all this energetic naturalism was created by pushing the silver from the back, and then refining the design on the front surface. Often these vessels would then be fitted with a silver liner to provide a smooth, easily cleaned interior.

Cup, Roman,
50–25 BCE, silver.

These techniques lend themselves better to some histor-ical styles than others. The effect of deep relief was particularly suited to the Baroque sense of drama, exemplified by the metal-work of the van Vianen family – celebrated Dutch gold- and silversmiths during the Dutch Golden Age in the seventeenth century. Their flamboyant silverwork is sometimes celebrated in the exotic paintings of luxury objects by Golden Age Dutch

Willem Kalf, *Still-life with a Silver Jug and a Porcelain Bowl*, 1655–60, oil on canvas.

Christiaen van
Vianen, rosewater
basin, 1635, silver.

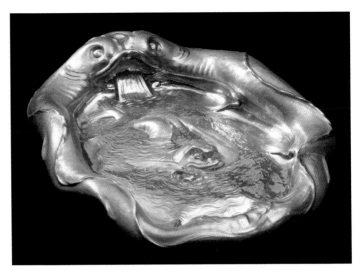

Miriam Hanid, *Union
Centrepiece*, 2013, silver.

painters. Experts in repoussé and chasing, the van Vianens are
credited with pioneering what has since the nineteenth century
been called the auricular style because of its rather disconcerting
resemblance to the fleshy forms of the ear (though at the time
it was more innocuously known as the 'whimsical' style). The
molten lines of this style, conjuring human faces, sea creatures
and waves, might suggest the pieces had been cast, but in fact

these vessels were raised and chased. This work did not only inspire Dutch Golden Age artists – recently, the Victoria and Albert Museum in London commissioned silversmith Miriam Hanid to create a piece as a tribute to Christiaen van Vianen's sinuous silver rosewater basin. Hanid's flowing centrepiece, with its chased decoration of waves, fins and tails, is as fluid as its inspiration. For now, the artist has arrested the movement of silver, but Hanid and van Vianen both remind us that silver is mutable and constantly refashioned.

Engraving

While chasing and repoussé move the malleable metal, engraving actually removes tiny slivers from the surface with a fine, chisel-like tool. This technique, too, can be employed to carry a narrative. London magistrate Sir Edmund Berry Godfrey proved his fortitude during the twin cataclysms of the Great Plague and Fire that scourged London in 1665 and 1666. His conduct was celebrated on a tankard engraved with scenes from the crises, Latin inscriptions about his efforts and the coats of arms of

Tankard, British, 1675/6, silver.

Paul de Lamerie
(engraving by William
Hogarth), salver,
1728–9, silver.

both Godfrey and the king. A more salutary view of London, a gracious backdrop to an engraved allegory, appears on a silver salver that was commissioned by the first prime minister of Great Britain, Sir Robert Walpole. Its maker was Paul de Lamerie, the most celebrated goldsmith in early eighteenth-century London. Lamerie belongs to the wave of talented Huguenots who emigrated to Britain in the late seventeenth century to avoid religious persecution on the Continent. In London, Lamerie used his extensive workshop of skilled craftsmen to execute the elaborate engraving for which he became renowned.

Engraving was frequently used for heraldic devices and to prominently announce a luxury object's ownership. This form of validation could be used to disconcerting effect. The silver Aedwen brooch is a piece of Anglo-Scandinavian jewellery from the eleventh century. The hammered silver disc is engraved with motifs of snakes and beasts, and an inscription in Old English on the back reads, 'Aedwen owns me, may the Lord own her. May the Lord curse him who takes me from her, unless she gives me of her own free will.'[5] The back of the brooch is damaged in a way that suggests the brooch was roughly torn from clothing

– perhaps a sign that the curse was less of a deterrent than its owner hoped.

As engraved lines can be calligraphic, the technique has been commonly used to 'write' on silver, making historic documents out of objects. The silversmith Paul Revere is more popularly known for his 'Midnight Ride' to warn the patriots of British troop movements at the start of the American Revolution. He belonged to the Sons of Liberty, a secret revolutionary society which commissioned him to make the iconic silver Sons of Liberty Bowl, engraved with the not-so-secret names of 92 prominent dissidents 'undaunted by the insolent menaces of villains in power'.

Piercing

Eighteenth-century puddings have been served on pierced trowels embellished with birds and foliage. Billows of incense have drifted to heaven through the pierced lids of silver censers.

Pierced lantern, Iran, early 19th century, silver and brass.

Temple pendant with filigree border, Kiev, 11–12th century, silver and niello.

Sugar, mustard and pepper have been shaken through the pierced caps of castors. And light has shone through Iranian lanterns whose pierced silver cases cast shadows of verse. The technique of piercing heightens the utilitarian to the sublime. Initially, piercing was a painstakingly slow process in which holes were drilled into the metal's surface, and patterns cut out with a wire saw. During the nineteenth century, the widespread use of the fly press for pierced work on baskets, cruets and castors transformed piercing into an industrial rather than an artisanal process, and brought such objects within the purchasing power of the middle classes.

Niello and oxidation

Silver is marred by tarnish, but niello is the application of black silver sulphide for decorative effect. The Roman naturalist Pliny wrote that niello was known to the Egyptians, but it became

Snuffbox, Russia, 1745–50, shell, silver, gilt-silver and niello.

much more common during the Roman period. The effect can be restrained or highly dramatic. A medieval silver pendant from Kiev portrays a mythical animal standing out in bright silver, against a dusky niello background. Russian silversmiths were virtuoso masters of niello, ingeniously transforming objects such as shells into silver-lidded snuffboxes. Gilded silver becomes a shimmering sky against which a grisly ship-wreck plays itself out in portentous shades of niello.

Bennett Kagenveama, cuff with silver overlay.

A related effect is the smoky patina of deliberately oxidized silver. A sulphur solution can be applied to create a charcoal-grey oxidized surface. Artists who have used this to great effect are the Hopi silversmiths of Arizona. The overlay technique they evolved was not a traditional one, but was created as a collaboration between the Museum of Northern Arizona and the Hopi tribe after the Second

Bowl, Near Eastern (Parthian), 1st century BCE, gilt silver and garnets.

World War to stimulate Hopi artisanship.[6] Traditional designs such as badger paws, corn stalks and spirit figures are traced onto sheet silver. These are cut out, and then soldered onto plain pieces of silver sheet. The areas behind the cut-outs are oxidized and often textured with punches. The raised surfaces are then highly polished to create dramatic chiaroscuro effects that are so striking that the overlay technique has been adapted by other Native American tribes.

Gilding

Gilding has been used to delight the eye far more often than to deceive. Although applying a thin wash of gold over a silver vessel might create the impression of a more valuable object, more frequently gilding has been used on select areas to contrast with the bright, white sheen of silver. A gilded bowl from the first century BCE illustrates the graphic possibilities of parcel (partial)

Michael Lloyd, 'Beech Leaves' beaker, 2014, raised and chased silver with gilt interior.

gilding. Gilded flowers studded with garnets are arranged within a silver net in a mix of Greek and Near Eastern motifs. Gilding could also be applied for a more pragmatic purpose: for centuries, the interiors of some food bowls and wine cups were gilded to protect them from tarnish and acids, creating a bright foil to the white exterior. Contemporary silversmiths sometimes emulate this to glowing effect.

Roman goldsmiths were skilled at the creation of hygienic food vessels, and at the delicate interplay between the warm and cool colours of gold and silver. Through the process of fire gilding, they were able to ensure a highly durable surface. A paste of mercury and gold was applied to the areas selected for gilding, and then the entire piece was heated to vaporize the mercury, leaving the gold chemically bonded to the silver ground. Fire gilding prolonged the life of the gilded surface, but not that of the unfortunate worker who inhaled the toxic mercury fumes. Despite the human and environmental cost, fire gilding

Georg Jensen, creamer from the 'Blossom' service, 1905, silver and ivory. The surface of this creamer is modulated by delicate hammer marks that scatter light.

continued to be the most common method until the nineteenth century, when it was supplanted by electro-gilding, which used an electrical current to plate the silver.

Finishing

To emphasize its lustre and reflectivity, silver has most often been polished to a high sheen, though sometimes, as in the case of niello, this is deliberately subverted. In pre-Industrial times, a smooth surface might indicate skilful use of a planishing hammer to even out any imperfections. A perfectly unblemished surface, though, wasn't always esteemed, especially following the Industrial Revolution. The Arts and Crafts movement in Europe and America was, in part, a reaction to the sterility of machined art objects. A feature of some Arts and Crafts-era silver is a surface deliberately imprinted by the mark of the hammer, and, by extension, the spirit of a human maker, rather than a machine.

Some cultures, too, prize the imperfect. The Japanese aesthetic of *wabi-sabi* finds beauty in the impermanent and imperfect. Although *wabi-sabi* in the arts is widely recognized in pottery, some Japanese silversmiths have also embraced the aesthetic, creating silverware characterized by asymmetry, and softly planished surfaces with visible irregular hammer marks.

Hiroshi Suzuki, *Earth-Reki III*, 2010, hammer-raised and chased fine silver 999.

4 Empire Building: Two Coins that Changed the World

It takes money to build an empire – not just money, of course, in the same way that it requires more than just fuel to stoke a fire. Culture, ideology, knowledge, need for land and resources all play their roles. But money is a basic ingredient. Money is what buys warships and equips soldiers. Money pays for the institutions of government that maintain empires. It finances civic programmes and architectural projects that promote a sense of shared identity. A common currency binds together countries and continents in trading partnerships and tax arrangements. To an empire's subjects, the images on the coins, which are handled daily, are a constant visual reminder of their rulers.

In the creation and maintenance of some of the most successful empires throughout history, those coins have been made of silver. The ownership and exploitation of silver mines bankrolled two empires that shaped the world we live in today. The legacy of the fifth-century BCE Athenian Empire was an idea of democracy, an intellectual revolution and philosophy that inspired literature and thought down the centuries, while the Spanish Empire two millennia later promoted international trade between the Americas, Europe and Asia, and a common world currency.

Athenian owl

Was there ever a coin as winsome as the Athenian owl? On the obverse (front) is the handsome, almond-eyed profile of Athena, patron goddess of the city of Athens. A goddess of both wisdom and war, her helmet is ornamented with an olive leaf (itself a symbol of her quick-wittedness in her move against Poseidon during a contest to claim Athens, in which she planted an olive tree), and her smile is gnomic – does she bring conflict or accord? On the reverse is her mascot: a small, standing owl, wide-eyed and flanked by an olive leaf, a crescent moon and the first three Greek letters for Athens and Athena. The owl was the quintessential coin of antiquity; it was the coin that passed through the hands of Plato and Socrates. The Athenian owl flitted far beyond the borders of its native Attica, populating the rest of the Greek mainland, the islands of the Aegean, the southeast Mediterranean, Palestine and the Middle East and parts of India. It was reproduced in Egypt. Today Athenian owls are highly desirable objects for collectors and museums, and are probably the most frequently forged type of antique coin. They embody an irresistible combination of beauty and narrative, bringing us the story of the ascendancy of Athens through silver.

There was nothing inevitable about the emergence of Athens as a colonial power, though it had several unique advantages that

Tetradrachm of Athens, *c.* 460–455 BCE, silver, obverse and reverse.

might have primed it for leadership. The roots of classical Greek culture extend back to a period of great transformation in the ninth and eighth centuries BCE, the era during which the Greek alphabet was developed, and the distinctive individual city states (*poleis*) were formed, each with its patron deity, temples and festivals. Many of these *poleis* were relatively small, covering only 50–100 km² (20–40 square miles), and some lacked an urban centre. Athens became a *polis* when the settlements of Attica agreed to a common political authority. At 2,650 km² (1,025 square miles) the state of Athens was considerable, and double the size of its commercial rival, Corinth. It was exceeded only by Sparta, its sometime rival, sometime ally, which controlled an area of 8,400 km² (3,245 square miles).

The patchwork of city states failed, however, to form a cohesive 'Greek' entity, preferring to look inwards for their local identities. *Poleis* separated by a short stroll could have completely different legal structures, and might swing between states of enmity and alliance. The political history of the *poleis* is a history of squabbles, skirmishes and complicated allegiances, punctuated by violent rivalry, such as that between Athens and Sparta. Diverse political and social systems coexisted in a small geographical area: Athens had a form of democracy by the sixth century BCE; Sparta juggled a complicated form of governance by kings, elders and an assembly consisting of all Spartan men over thirty years old.[1] Neither were the *poleis* united in currency, as the larger city states minted their own coins. Athens minted silver coinage from the mid-sixth century BCE, mainly from non-local silver.

The Laurion mines

This situation changed dramatically around 520 BCE, with the discovery of rich silver deposits at Laurion in south Attica – not that the Athenian state was the first to exploit these prodigious resources; they were known as far back as the Bronze Age. The landscape of Laurion displays plenty of visual cues about potential riches which could attract attention; the hills contain not just silver and lead, but zinc, iron, copper and gold. Unweathered minerals might be revealed after a tree is uprooted. In places, the hillsides are visibly striated. Geologically, layers of calcite alternate with layers of schist. Hydrothermal fluids deposited silver-rich minerals in the areas of contact between the two main rock types.[2]

The Bronze Age mining at Laurion was open-cast mining. Gradually miners followed the mineral trail from the surface into the hillside (drift mining). It is still unclear exactly what triggered the bonanza that was to enrich the Athenian state. By the time the Athenians discovered the resource-rich land, the surface had been scraped clean of the most valuable metals. Probably it is safest to assume that those investing in mining were 'hunting elephants in elephant country'. (Following the same principle, the Laurion mines were returned to later by the Romans.) Vertical shafts were dug deeper and new contact zones discovered, attracting investment by an increasingly money-minded Athenian elite. Significantly, as slavery was entrenched under the *polis* system, there existed a large slave labour force that could be mobilized in the search for and extraction of precious metals.

The mines at Laurion belonged to the state of Athens, but private citizens could buy a commission for a specified number of years – usually between three and ten – ensuring a steady stream of revenue for Athens. Obviously such a system held risks for entrepreneurs, as there was no guarantee of striking rich. Evidently, many found the gamble worthwhile, as the Laurion landscape today, from the hills down to the coast, is riddled with evidence of shafts, galleries, slag heaps and the remains of processing plants. Although the honeycomb of workings and workshops existing cheek-by-jowl might suggest this activity was haphazard, it was in fact tightly regulated. Records form the fourth century BCE reveal the different spheres of activity that offered attractive opportunities for investment – from discovery, to creating and operating a mine, to processing and refining.[3] Opening a mine did not demand a great deal of capital, but it did require the ownership of slaves or the ability to hire labour. Much more capital was needed to invest in processing and refining plants and the hiring of skilled labour, though this was obviously a less risky enterprise. Not surprisingly, the affluent generally engaged in the latter.

Laurion was, in essence, an industrial-scale slave camp where conditions were predictably brutish. At times, the hillsides were worked by possibly 20,000 to 30,000 slaves – numbers close to the free population of Athens[4] – referred to in ownership inscriptions as 'human cattle'. Most of these slaves were not convicts but prisoners of war, slavery being a highly profitable by-product of the ceaseless warfare that characterized the Aegean world.

Writing slightly later, the historian Xenophon recorded that the largest mining lease investor at Laurion owned 1,000 slaves.[5] Such large numbers were required because of the fragmented pattern of small-scale leases, and the arduous nature of the manual labour; three teams of six men might have taken two years to dig to a depth of 100 metres (330 ft).[6] They laboured in cramped galleries that generally could accommodate one stooped person, with basic tools such as hammers, picks and an oil lamp that burned for a ten-hour shift. Rock faces were broken up by fire-setting – a dangerous method involving heating the rock and then dousing it with liquid to fracture the surface. After a preliminary sorting of ore underground, it was sent to be 'dressed' in surface workshops. Like the mines, these workshops were individual investments. Here the ore was hammered by hand, washed to remove the dross, pressed into patties with human dung (a plentiful commodity) and dried. At this point, it was ready for smelting to isolate the silver – a more centralized activity, as it required skilled labour. It is likely that Athens also received income from the furnace operations, perhaps in the form of a tax on refined silver.[7] So Laurion was two cities: one below, and the other above ground, and in both the conditions were nasty and hazardous. The air would have been pungent with the rotten-egg stench of sulphur from the furnaces. Worse still, a by-product of the process to separate silver from lead is lead-oxide – a fine white powder that drifted into the atmosphere. Over time, millions of tonnes of toxic ash were released into the environment.

Following the initial extraction of rich silver ore around 520 BCE, the Athenian mint began mass production of the owl tetradrachm (four drachmas), which replaced the smaller traditional didrachm (two drachmas) as the major denominational weight unit.[8] These owls were struck in their millions, but the industrial scale of production was not achieved by industrial methods. The mint workers were probably public slaves owned by the state.[9] They hammered out each owl by hand, placing a silver blank between an engraved anvil and punch and compressing it with a mighty blow. In the mid-fifth century, the tetradrachm was worth four days' pay for a skilled labourer, though it is doubtful any slave ever owned an owl produced by his labour.

It was the silver from the Laurion mines that was to precipitate a sea change in Athenian affairs in relation to the rest of Greece and beyond, notably to the rapidly expanding Persian Empire to the east.

The Persian wars

Greece's extensive coastline and thousands of islands orientated its inhabitants seawards, leading to the founding of cities and colonies stretching around the Mediterranean coast to the western fringes of Asia Minor. Wealthy Corinth had an extensive trade network and dozens of colonies. Athens had displayed expansionist ambitions during the seventh and sixth centuries, with the planting of a strategic outpost near Troy on the far coast of the Aegean Sea, and the annexation of the nearby island of Salamis, secured for its excellent harbour.[10] The rise of Athens, though, can only be defined in relation to the Persian Empire.

While the system of city states did much to divide the Greek people, it was the outside, Persian influence that helped to forge a sense of Greek identity. By the beginning of the Athenian exploitation of Laurion, the Persian Empire seemed unquenchable in its thirst for territory. From its base on the Iranian Plateau, it had conquered Mesopotamia, the eastern Mediterranean and parts of Egypt. With its vast military, sophisticated infrastructure and administrative flair, it became a significant threat in the Aegean, even though its heartland was 3,200 km (2,000 miles) and a three-month march to the east. The Greek city states nibbling at its western border were swiftly brought under Persian control, and forced to pay tribute to Persia.

In 499 BCE an Athens-supported rebellion among some of these states resulted in the dispatch of Persian troops to Attica. It ought to have been an easy victory for the Persians, but in 490 BCE, on the boggy plains of Marathon just outside Athens, the Persian infantry were ignominiously slaughtered. A decade later the Persians under King Xerxes the Great returned to Greece with intimidating land and naval forces. One after another, the Greek city states capitulated, and when Persian troops marched into Athens, they found the city eerily deserted. The inhabitants had fled to the neighbouring island of Salamis, some leaving their Athenian owls behind in their haste – a hoard bearing burn marks was discovered on the Acropolis in 1886, in a layer of scorched earth dating back to the Persian destruction of the area.

On land the Greeks had been routed by an army so rapacious that the historian Herodotus claimed, 'What body of water did his forces not drink dry except for the greatest of rivers?'[11] At sea, though, thanks to the silver of Laurion, it was a different matter. In anticipation of Persian reprisals after Marathon, seaward-focused Athens had concentrated on building its navy, funded by revenue from the Laurion mines. Having lured Persian ships into the straits beside Salamis, the Athenian navy inflicted a humiliating defeat on the enemy fleet. The tide of the Persian Wars turned, and King Xerxes returned home in disgust, having had his remaining sea captains executed. The battle was commemorated by the Greek dramatist Aeschylus in his tragedy *The Persians*, in which the Athenian advantage is alluded to: 'Of silver they possess a veritable fountain, a treasure chest in their soil.'[12]

The rise of Athens

While the naval victory at the Battle of Salamis earned Athens great acclaim, the threat of renewed Persian aggression continued to haunt the Aegean. In 477 BCE most of the Aegean islands banded together to form the Delian League. The main advantage of League membership was protection by its navy, to which large cities like Athens contributed ships and crews, while the others paid annual dues. Funds were held centrally in the Treasury at the sacred isle of Delos in the Cyclades. Ostensibly the Delian League united the city states against a common enemy; all too soon, though, any pretence of solidarity disintegrated into the factionalism and self-interest that seemed the default position of the *poleis*. Athens jockeyed for a position of superiority, annexing a Persian trading post in the north Aegean for itself. It prevented disgruntled members from exiting the League, and ruthlessly crushed discord. The removal of the League's treasury from Delos to Athens in 454/3 BCE was one more wanton flouting of unity in a succession of Athenian bids for power.

Athens had become an empire – albeit a small and short-lived one. At its height, it ruled 179 states encompassing about two million Greeks, and was the dominant naval power in the Mediterranean from Sicily to Egypt, to the Black Sea. Although miniscule in relation to Persia, it behaved like an empire, exacting tribute from its subjects in the form of monetary payments and decreeing the use of Athenian coinage throughout its lands. By the middle of the fifth century BCE, the Athenian mint had ramped up production to new heights, with standardized owls flowing in abundance as an international currency throughout the Greek world.

Preferred by merchants as a reliable currency well beyond the borders of the Athenian Empire, this coinage also flowed right back into Athens through tribute and trade, vastly enriching the city. Like all image-conscious empires, Athens used some of this wealth to embark upon a public building programme, the jewel of which was the Parthenon on the Acropolis, begun in 447 BCE. Recent research has proposed that the attic of the Parthenon,

The Parthenon,
Athens, from the
south.

which was the size of three tennis courts, held the city's cash
reserves.[13] One decree records the transfer of 3,000 talents of
silver to the Acropolis. As there were 1,500 tetradrachms to the
talent, this transaction could have consisted of 4.5 million silver
coins. Other ancient sources mention the stockpiling of 10,000
talents. No wonder Aeschylus commented on Athens' 'veritable
fountain' of silver.

Architecturally, the Parthenon might have been a symbol of enlightenment and an embodiment of well-proportioned beauty. Its famous frieze depicting mortals and gods could have been intended to exemplify order and harmony. If the city's massive silver reserves were indeed stored in its attic, under the protection of Athena herself, this might be understood as a policy of prudent government. But the reverse side of Athens' silver-fuelled empire was the exploitation of slaves, the annexing of the Delian League's treasury, coercion and extortion of former allies, the subjugation of territories, administrative abuse and corruption. Predictably, some of the subject states revolted, and these revolts were countered by bloody reprisals. For instance, when the island of Melos resisted absorption into the Athenian Empire in 416 BCE, all the adult males were slaughtered, and the women and children enslaved.

As with most empires, though, territorial ambitions over-reached the capacity of Athens to maintain stability. It was, in the end, the Athenians' traditional enemy, Sparta, assisted by Persian funds, which brought down the city state. Ironically, it was a naval defeat that brought Athens to its knees. Its fleet destroyed, the city was besieged and starved into capitulation in 404 BCE.

Despite its imperial ambitions being ended, the culture Athens nurtured continued to flourish. Politically, Athens had offered a model for democracy, albeit one that limited citizenship and excluded women. Intellectually, it fostered a climate of debate and self-examination that underpinned the astounding achievements of classical Greek philosophy. It championed an openness that allowed theatre to function as social critique. It was the cradle of what we have come to recognize as Hellenist culture. Under its patron Athena, goddess of wisdom, it was, as claimed by the Athenian politician Pericles, 'the school of Hellas'.[14]

Pieces of eight

'Pieces of eight! Pieces of eight!' squawks Captain Flint, Long John Silver's parrot. The *peso de ocho reales*, the Spanish 'piece of eight', was first minted in the 1570s, but *Treasure Island* was set in the eighteenth century, and written by a Scotsman in the nineteenth. How did this Spanish coin, so beloved of pirates and parrots, maintain its dominance for so long?[15] And how did it command a terrain that stretched from the Americas across Europe, into Africa and across Asia to the Far East? Around 1600 it was legal tender virtually anywhere (although it was eventually supplanted in the nineteenth century by the British pound sterling).[16] Like the Athenian owl, its story is embroiled in the creation and collapse of another unlikely empire that found itself in possession of a prodigious silver mine.

Just like Athens, Spain did not seem unquestionably destined to become an empire. It was almost a backwater of Europe, and attained much of its European territory through dynastic succession, rather than ruthless expansion.[17] Its American territory came about almost by accident. When the Genoese merchant and sea captain Christopher Columbus approached Ferdinand and Isabella to sponsor his sea journey to Asia (not America), and convert the Asian Muslims to Christianity for the glory of the Spanish crown, he had actually already approached the Portuguese, French, English and Spanish courts before. It was possibly a surprise when Ferdinand and Isabella at last said yes. The venture may have been proposed as a bid for territory and the suppression of Islam, but it was also a quest for gold. Columbus

Peso de ocho reales, Potosí mint, 1613–16, silver, obverse and reverse.

eventually made four journeys to the New World, and died believing he had genuinely reached Asia. In the West Indies, he had discovered disease, hardship and rebellion, but little gold.

A few decades after Columbus's death in 1506, the paltry gold mines of the Caribbean were worked out. The next bid in the search for El Dorado, the legendary land of gold, came when Hernán Cortés invaded Mexico. To the subjugated Aztecs, whose empire was so ferociously overthrown, the conquistadores seemed like swollen-bodied pigs in their greed for gold.[18] This initial stage of empire-building was not part of any systematic plan by the Spanish monarchy to strategically expand its sphere of influence. In fact, Cortés's expedition, with only six hundred men and sixteen horses, was made without royal approval. In a stunt worthy of a Monty Python sketch, Cortés was forced to interrupt his kidnapping of the Aztec emperor Montezuma in order to waylay his own arrest for disobeying the crown.[19] The gold recovered from such exploits, though, fuelled further Spanish ambition, and Peru appeared to glitter with promise. The geographically large and distant Inca Empire, ruled over by an emperor revered as a descendent of the Sun God, was brought down by a small force of men led by an illiterate former swineherd and a fugitive who had escaped to America to evade Spanish law.[20] In 1532 the unfortunate Inca ruler Atahualpa was captured, and, just as Montezuma had, attempted to bargain for his life with treasure – a room filled once with gold and twice with silver. The Spaniards accepted the ransom, but garrotted him anyway. These early foraging expeditions, ill-organized and often fractured by internal discord, failed to discover El Dorado. They did, however, stumble upon the precious lodes of silver that, in financing Spain's own empire, altered the balance of European power. By the mid-sixteenth century, rich silver deposits had been discovered in the western Sierra Madre range in Mexico. But the real jackpot, the silver mountain, was struck at Potosí in the Andean mountains of modern Bolivia.

Potosí and the Cerro Rico

The Andean Indians had a robust three-millennia-long trad-ition of metallurgy, and had been working in gold, silver and bronze far longer than the Mexican peoples. Most of their metal artefacts seem to have been for religious purposes, rather than as a store of wealth. At the silver mines at Porco, Andean miners dug out silver ore with deer horns, and smelted it in small ovens.[21] Sometimes the indigenous peoples, hoping to gain favour with their new overlords, revealed sources of silver to the Spaniards; more often, though, they were coerced. The other-worldly conical mountain that loomed above Potosí was long considered a sacred place, and some native silver might have been discovered during visits to hillside shrines. Accord-ing to one legend promulgated by colonists, native miners were admonished by a booming voice when they first tried to extract silver: 'Take no silver from this hill for it is for other owners.'[22] Gleefully, the Spaniards justified their exploitation of the Cerro Rico, the rich mountain, as a fulfilment of this prophecy.

As word of the silver mountain spread, Spaniards as well as some Indians flocked to Potosí. During this early phase of

'A Section of a Silver Mine in Potosí', 19th-century engraving for the *Universal Magazine* (London).

A SECTION of a SILVER MINE in POTOSI, and the Manner of WORKING it.

Engrav'd for the Universal Magazine, for J. Hinton at the Kings Arms, in St Pauls Church Yard London.

the mine's development from 1545, extraction was fairly simple. The efficacy of the supernatural prohibition against mining, or lack of local awareness of the deposits, had resulted in the survival of very rich oxidized ores, including native silver. Once the surface ores had been exhausted, though, more costly mining methods were needed. Tunnels began to burrow deeper into the hillsides, where the ore was less rich. Local smelting methods were inefficient for the vast volumes of lower-grade ore the Spaniards needed to process, and new technologies were needed to maintain production levels. In Mexico, a method for refining low-grade ore had been implemented in the 1550s. The environmentally ruinous 'patio process' involved mixing crushed ore with salt and mercury, supplied from the great mercury mines at Huancavelica. The addition of a reagent prompted a chemical reaction in which the silver formed an amalgam with mercury. Heating the amalgam released the mercury, leaving refined silver. The adoption of this process at Potosí in the 1570s had the effect of opening a spigot – silver output almost tripled during the 1580s and 1590s. At its height in the 1590s, the annual output was 200,000 kg (440,000 lb).

Novel technology increased the need for human labour. But Potosí, at over 4,000 metres (13,000 ft) above sea level and on the fringe of a desert, was a bleak and inhospitable place, hardly attractive to the Indians settled in agricultural villages in the valleys. Successive Spanish monarchs had repeatedly forbidden the enslavement of the empire's Indian subjects, and a Papal Bull expressly condemned the practice. But a devious workaround to this inconvenient law was the adaptation of the Incan institution of the *mita*. As a means of fulfilling their tribute obligations, native tribes had traditionally supplied workers for imperial projects on short-term rotation. The Spanish version of the *mita*, implemented at Potosí from 1573, epitomized the dark underbelly of empire. Far more exigent than the Incan *mita*, the Spanish system cast a net extending hundreds of kilometres from the rich mountain in order to round up half the workforce needed for the mines. The far-flung provinces were forced to conscript one-seventh of their eligible male population between the ages of eighteen and fifty each year as tribute *mitayos* (forced participants in the *mita*). Some of these *mitayos* had a trek of 1,000 km (600 miles) and needed several weeks just to reach Potosí before their year of service began. Some made the journey alone, unsure whether they would ever return to their homes. Others brought their families with them; wives could perform menial labour, while children could scavenge.

Predictably, given this stream of cheap labour, conditions in the mines were lamentable, with scant consideration paid to safety. Teams stayed underground for a week, working by candlelight in cramped and tortuous tunnels. Malnourished, continually at risk from accidents and cave-ins, frequently beaten and breathing silica-laden dust, the *mitayos* sucked on coca leaves wadded against their cheeks to dull hunger and tamp down anxiety and fatigue. The wages they received were barely subsistence. Some found themselves living on the street as the population of Potosí swelled to over 100,000. Most *mitayos* spent much of their year cold, hungry and afraid. For those who survived, there was no provision for their journey home, and some had to remain in the city, often hiring themselves out as free labourers capable of earning higher wages as experienced miners. Others begged their way back to their villages, knowing that they would have to repeat their ordeal within the decade, and that any given year their brothers, fathers and sons would experience the same. Lives as well as silver flowed out of the rich mountain. 'Every peso coin minted in Potosí has cost the life of ten Indians who have died in the depths of the mines', claimed the Augustinian monk Fray Antonio de la Calancha.[23] Throughout the colonized areas of Mexico, Central America and Peru, disease and privation had by 1570 decimated the native population by 80 per cent, and numbers continued to drop until the middle of the next century.[24] There were no Spanish laws against the enslavement of Africans, and consequently tens of thousands of African slaves were shipped to Peru to replace native workers.

Meanwhile, Potosí grew into a vibrant, multicultural and rich city, adorned with magnificent ecclesiastical and domestic architecture. The barren land around the city might produce only potatoes and alfalfa, but soon its wealthier inhabitants were supplied with melons and lemons, spices and sugar, fine wines and brandies, meat from Buenos Aires, luxuries from France and the Netherlands, porcelain and silk from Asia, diamonds from Ceylon, carpets from Persia and crystal from Venice – all hauled by mule or llama along mountain paths from the coast.[25] Potosí silver also purchased the African slaves who worked in the mines and almost every home of substance in the district. The city's coat of arms gloated, 'I am rich Potosí, treasure of the world, king of all mountains and envy of kings.' The King of Spain granted it the title of Villa Imperial (Imperial City) in 1559. Its reputation spread so fast that soon the expression *vale un Potosí* (worth a Potosí – that is, priceless) entered the Spanish language. Cervantes has Don Quixote tell the faithful Sancho Panza, 'If I could only compensate you for what you deserve, the mines of Potosí would not suffice.'

Of course, it was not only individual mine owners and successful miners who gorged upon the profits from the mines. The Viceroyalty of Peru, established by Spain in 1542 to govern its American domain, reaped enormous profits from the silver that flowed to its coastal capital, Lima. Staggering tonnages of almost-pure silver was loaded onto ships bound eventually for Seville, but plentiful amounts of silver also remained in cities such as Potosí and Lima. The Church became a major patron for both native and European silversmiths who crafted fine Baroque altar fronts, crosses and statues. The early historian of Potosí, Bartolomé Arzáns de Orsúa y Vela, compared the ripe ostentation of one Potosí church interior to 'a blooming jungle, with a great number of braziers of the purest Cerro silver, amber from Florida, precious aromas from Araby, silver pomanders filled with simmering fragrances activated by dancing flames'.[26] In the popular imagination, the streets of these cities were paved with silver – a fancy that became literally true during state celebrations such as a festival to welcome the new viceroy to Lima in 1648.[27]

Paying for much of this decadence and devotion were the *pesos de ocho reales*, struck at Potosí's royal mint. Once refined,

Altar plaque with archangel, Peru, *c.* 1640–50, silver. Lavish silver ornaments embellished churches in Peruvian cities.

the silver was taken to the treasury office, where it was assayed, registered and taxed. At that point, the silver destined for coinage was taken to the mint, founded in 1575. In its earliest years of operation, the mint was staffed by indigenous *mitayos* and African slaves. As production ramped up, the mint's rooms were leased to subcontractors who installed slaves from Angola and the Congo. Just as in Athens, these slaves hammered by hand the millions of silver coins which lubricated an economy that ingested slaves as a commodity. By 1640 over 150 slaves worked in the Potosí mint, coining five million pieces of eight each year.[28]

5   Rivers of Silver from the New World to the Middle Kingdom

Potosí's chronicler Bartolomé Arzáns de Orsúa y Vela had a gift for an apt turn of phrase, observing that silver in the Viceroyalty of Peru was 'a kind of River upon which all useful and necessary things sailed and were transported'.[1] It might have pooled in the churches and streets of Lima and Potosí, but plenty of it rushed in cataracts towards Asia and Spain. It was borne on the galleons that made their perilous journeys across the Atlantic to Europe, and across the Pacific to the Philippines.

Treasure fleets to Spain

In the first arduous stage of its journey, the silver was loaded onto mules and llamas at Potosí and hauled along bandit-infested mountain paths to the port of Arica in present-day Chile, the driest settlement on earth. A train might comprise thousands of beasts of burden, accompanied by hundreds of Indians and armed Spanish guards. In the arid dunes around Arica, many llamas dropped dead of thirst and were abandoned to mummify in the dry air. Once they reached port, the tonnes of silver bars and coins were loaded onto ships that would sail north along the coast to Panama, navigating an unprotected stretch of water that proved a playground for pirates. The most infamous was Francis Drake, waging his war against Catholicism in the name of Protestantism (and personal profit) by pilfering the Spanish Crown's most precious resource. In 1579 Drake pillaged his way up the coast from Arica, capturing merchant ships stocked with so much silver that they used some as ballast. He helped himself to registered and unregistered – that is, smuggled – silver, stealing from the Spanish as well as from other pirates, and in the process stocking up on the salt pork, hams and wine that also made up the purloined cargoes.[2] Drake's temerity stunned the Spanish into forming an armada to protect the ships shuttling silver and supplies between Panama, Arica and Lima's port of Callao.

Once the Potosí silver reached Panama, the entire cargo was unpacked and loaded onto mule trains. The overland route across the isthmus through the dismal, disease-ridden valley of the Chagres River came to be known as the 'Road of the Crosses', after the graves of hundreds of men lost to disease and accidents. The port of Nombre de Dios, their destination on the Caribbean coast, was surrounded by swampland and thorny brush infested by pestilence and pirates – including, once again, Francis Drake. Pirate raiding parties relieved the Spanish of silver and slaves, as well as, on one embarrassing occasion, a shipload of official and personal correspondence from Spain.[3]

For its next leg of the journey, the treasure received better protection. From the early years of its involvement in the Americas, the Spanish Crown had forbidden journeys by single ships and instead organized fleets that sailed under the protection of heavily armed galleons, funded by the ships' owners. From Nombre de Dios, the convoy sailed north to rendezvous with the fleet transporting silver from Spain's mines in Mexico. Then the entire fortified floating treasure trove made its cautious way back across the Atlantic towards Seville. As it approached Spain, it was met by warships that escorted it home. Although Seville was 80 km (50 miles) from the coast up the Guadalquivir River, it had been chosen by Isabella and Ferdinand as the city through which all trade between Spain and America was conducted. Seville's *Casa de Contratación* (house of commerce) registered and levied taxes on imports of silver. For the Spanish Empire, these taxes were the lifeblood that sustained the edifice. Seville was the heart that pumped profits equalling up to 40 per cent of the value of the imports.[4] The most onerous was the *quinto*, or Royal Fifth, the flat tax of 20 per cent claimed by the Crown. In the

Anthonis Mor
(manner of), *Philip II
of Spain*, 1560–1625,
oil on canvas.

usual manner of taxation, though, many additional indirect taxes were levied. It was little wonder that, despite the dangers, smuggling was rife. Partly for this reason, it is difficult to ascertain exactly how much silver flowed from American mines into Spain. Between 1500 and 1650, 16,000 tonnes were officially processed through Seville.[5]

How Spain was to use this silver was apparent from the first shipments that arrived after the earliest conquests. Silver from Peru started flowing back to Spain with Atahualpa's ransom, in the form of ingots and Inca statues of almost life-size animals, plants and humans. However, these sophisticated works of art were put to the immediate purpose of war in the funding of King Charles V's invasion of Tunis and the battle against the Turkish army. The rise of the Ottoman Empire during the fourteenth and fifteenth centuries was perceived not just as a threat to Spanish territory but as an affront to the Christian faith. Bankrolling military expenditure with American silver was a Spanish strategy that was to play out in full during the remainder of the empire's existence.

The reign of Charles V's son, Philip II, from 1556 to 1598, coincided with the surge in silver production from Spain's American mines. Yet despite the staggering quantities of bullion processed through Seville, most of it simply passed through Spain into hands far beyond its borders. It was as though the silver really was a river that flowed straight through Spanish fingers. Not all of this silver was used by Spain as capital with which to improve agriculture, develop infrastructure or advance industry. Some lavish residences – notably the splendid royal palace El Escorial – were constructed, and literature and the visual arts enjoyed a renaissance that clearly benefitted from the economic boom. But the majority of Spain's new income was spent on warfare and the repayment of debt. At the time of Philip II's accession, Spain had territories in Italy, Burgundy, the Netherlands, the Americas and North Africa. It collected taxes and revenues and derived enormous income, especially from the Americas, but quelling uprisings and addressing the persistent threat of rebellion drained the treasury.

Like his father, Philip II's agenda was to strengthen and protect Spanish territory, as well as suppress Protestantism and Islam, the twin affronts to the Catholic faith. The *quinto* contributed to the enormous outlay in military expenditure during the 1570s and 1580s, but it proved inadequate to the empire's perceived needs. As Philip II's reign wore on, battling the Protestant rebellion in the Netherlands proved particularly draining. The humiliating defeat of the Armada in 1588 – a spectacularly expensive campaign – bruised Spain's purse and prestige. Its enemies were encouraged to foment further unrest that proved costly to contain.

An additional evisceration of Spain's revenue came in the way of financing its debt. Charles V's reign had been one of almost-constant warfare. The loans he received from European bankers carried high interest rates, and his son inherited not just territory but debt. Despite the surge in revenues during Philip's reign, the crippling cost of empire forced him to turn again to bankers in Genoa, Antwerp and Augsburg. While 'rich Potosí' was crowing over its wealth and paving its streets in silver, Spain was lamenting its penury and declared bankruptcy in 1557, 1575, 1596 and on several occasions in the seventeenth century.

Another scourge was one endured not just in Spain but throughout the world. Like a jolt of sugar in the bloodstream triggering a hypoglycaemic crash, the huge amounts of silver rushing into the interconnected economy helped spark a widespread economic collapse. The more that silver gushed from the Americas, the more it lost its value and purchasing power. The Spanish monarchs discovered a disconcerting economic principle – abundance can be an affliction as well as a blessing, as monetary expansion drives prices higher.[6] With surging inflation, the cost of living doubled over the second half of the seventeenth century. The Spanish Crown claimed the same amount of taxes in silver, but it was worth so much less. Spain's debt-addled economy continued to sour, the peasantry went hungry and uprisings of angry citizens erupted across Europe. The ramifications of this 'price revolution' were felt across the globe.

Silver to China

The merchants of Seville complained in 1626 that 'the foreigners are rich; and Spain, instead of being as a mother to her sons has ended up as a foster-mother, enriching outsiders and neglecting her own.'[7] Usurious European bankers might have fallen into this category, but Spanish discontent was probably directed further eastwards, specifically to China. The 'silver sink' of China was the final destination of between one-third and half of all silver mined in the Americas.[8] China seemed the ocean into which so many rivers of silver inexorably flowed. China had a long and rich tradition of silver-working, both for religious and domestic uses. But most of the silver imported into the country satisfied an entirely different need.

The drainage of silver into China was long explained in terms of a trade deficit. Europeans craved eastern luxuries such as silk, spices, porcelain and tea, but the self-sufficient Chinese wanted nothing that the West could offer, resulting in the emptying of the European bank account – until the nineteenth century, that is, when European traders foisted opium on China in exchange for bales of silk and chests of tea. This viewpoint, though, neglects the Chinese thirst for silver as a powerful current running in the opposite direction.[9]

Unlike European countries, which had predominantly used precious metals for currency, China had evolved a different system of finance. Between the eighth and fifth centuries BCE, bronze (an alloy of copper and tin) had been cast into the shapes of spades and knives to be used as currency.[10] The later, familiar square-holed bronze coins were in circulation for several centuries before the birth of Christ. A shortage of copper and the low intrinsic value of each coin encouraged the official implementation of paper money in the twelfth century. Hundreds of years before paper currencies were adopted in Europe, China introduced the notion of 'fiat' money (whose value lies in government guarantee), as opposed to commodity money (whose value lies in the worth of the precious metal). One of the problems with paper money, though, is that injudicious use of the printing press almost inevitably leads to inflation – a consequence discovered by the Song emperors in their efforts to fund wars against the Mongols. Over the following centuries, succeeding dynasties attempted to enforce the use of paper currency despite ruinous cycles of inflation. Eventually, lack of confidence in the paper system led the Ming government (1368–1644) to accept un-coined silver for tax payments.

China did have its own silver mines in Yunnan, in southwest China, but these could not supply the needs of a country which made up one-quarter of the world's population, with cities one million-strong by the seventeenth century.[11] Nor could the Japanese silver mines of Iwami Ginzan keep up with demand. Another spur to the 'silverization' of the economy had come with the implementation of the 'Single Whip' tax in the 1570s, which consolidated separate taxes into one payment – to be made in silver. The value of silver began to soar, and canny merchants seized the opportunity for some highly profitable arbitrage. In China during the Ming Dynasty, silver was worth double its value anywhere else. For instance, at the southern port of Canton, gold's value in relation to silver was 1:7. In Spain, the ratio was 1:14.[12] In other words, gold could be purchased for less silver in China, and traded for more silver in Europe, leading to a

Belt plaque in the shape of a crouching horse, China, 3rd–1st century BCE, silver.

Dish shaped as a leaf, China, late 7th to early 8th century (Tang Dynasty), silver with parcel gilding.

drain of China's gold by those astute and mobile enough to make the trade.

Such arbitrage, obviously, relied on a certain freedom of movement – an openness that was hardly characterized by the restrictive trade policies of Ming-era China. How did these early arbitrageurs gain access to their markets? The answer lies largely in the entrepôt of Manila.

The region was a marketplace long before Spanish interest in the Philippines. Strategically situated between China, Japan and the Spice Islands, it had long attracted Chinese and Indonesian merchants who traded in gold, porcelain and guns.[13] But it had an unlikely beginning as Spain's international trading post. In 1519, in a bid to reach the Spice Islands, the Portuguese explorer Ferdinand Magellan embarked from Seville on a journey across the Atlantic. After navigating the passage that became the Strait of Magellan, he crossed into the Pacific, eventually reaching the islands that were later to be named after Philip II. Magellan died there in a confrontation with the natives, but a battered remnant of his crew returned to Spain, having barely survived the first circumnavigation of the globe. It took until the 1650s for Philip II to authorize the founding of a trading colony in the Philippines. The city of Manila was founded as a Spanish colony in 1571 – the birth date of global trade, according to prominent historians of the silver trade in China.[14]

A far-flung outpost of the Spanish Empire, administered by the Spanish Viceroyalty of New Spain, Manila was divided into separate enclaves. One was the official 'European' town and administrative heart of the colony. Steamy, malaria-ridden Manila was an unattractive destination for European colonists, and for decades European residents never numbered more than a few hundred administrators, traders and Roman Catholic missionaries. And then there was the Parián, a teeming Chinese settlement of thousands established by the colonists in 1583 to corral Chinese merchants and immigrants. A curfew ensured that the Chinese vacated the walled colonial city by nightfall. The Parián was a warren of warehouses, teahouses, shops and residences that formed the thrumming, mercantile heart of Manila. By the mid-seventeenth century, Manila's population had surged to 42,000 residents, with its 15,000 Chinese outnumbering the Spanish 2:1. In addition to 20,000 Filipinos, there were also thousands of Japanese.[15]

From the start, the relationship between the Chinese and colonists was one of mutual dependence and distrust. Manila was a stepping stone in the Pacific whose sole purpose – in the minds of the foreign parties – was as a centre for the exchange of Chinese goods and Spanish silver. Somehow, two vastly different cultures had to manage their collision and collusion. The colonists depended on the Chinese as buyers of their silver, but also as their craftsmen, cooks, labourers, fishermen, dentists, bakers and household servants.[16] Yet they resented their numbers, and the acuity of their negotiations, and possibly their intractability in the face of efforts to convert them to Christianity. The colonists repeatedly called for the expulsion of the Chinese, imposed onerous taxes on their transactions and brutally suppressed their protests, but in the end, they couldn't even bake palatable bread without them, as both bakers and wheat flour came from China.[17]

Although the Chinese had travelled to the Philippines seven centuries before Spanish colonization, it was Spanish Manila that became the largest offshore port for Chinese goods. Most of these goods came from the southern province of Fujian, through a small number of ports such as Yuegang. Rocky, infertile Fujian had long looked to the sea for its sustenance, and had early embraced maritime trade. Before the founding of Manila, Yuegang might have sent one or two small ships to the Philippines each year, but by the 1580s it was sending at least twenty large junks laden with silk, porcelain, cotton, ivory, precious stones, lacquerware and furniture.[18] The small traders who rented space on the junk from wealthy ship owners had a relatively short voyage of ten days sailing. But they still had to risk the long-familiar perils of shipwreck and pirates. After docking, they paid customs duty on their cargoes and found agents to sell their goods on to the Spanish in exchange for silver. Quite quickly, the Chinese learned how to exploit the tastes of overseas markets. Initially, they exported silk in bolts; soon they acquired samples of Spanish dress and began to supply finished garments perfectly attuned to fashionable European tastes. While being taxed as non-Christians, they supplied sumptuous silk vestments for the California missions, gilded Virgins and marble statues of the Baby Jesus.[19]

The Manila galleons

In addition to their cultural differences, there was another reason for the tension between the colonists and the Chinese – much of the silver was unregistered, and therefore contraband. During the seventeenth century, 2 million pesos, or about 50 tonnes of silver, were shipped annually across the Pacific from American mines.[20] Reporting quantities with accuracy is hindered by the difficulty of accounting for smuggled silver, but in some years as much as 5 million pesos may have been shipped to the Philippines.[21] The prodigious quantity made the Spaniards' loss of tax revenue all the more galling.

Both Peruvian and Mexican silver flowed along the Pacific route out of the Americas to Asia. Direct trade between Peru and the Philippines was, for most of the period, prohibited by

Panel, made in China for export, 17th century, embroidered silk and gilt
paper-wrapped thread.

Mater Dolorosa (Mourning Virgin), Qing Dynasty, 18th century, Sino-Spanish, wood with pigments, gilding, ivory and silver.

the Spanish Crown, though large quantities were smuggled out through the 'back door' of the Andes to Buenos Aires on the Atlantic coast. Officially, Acapulco, a deep and safe harbour on Mexico's Pacific coast, was the single point of exit and entry for silver and goods traded between the Americas and Manila. The story of the Manila galleons is one of astonishing longevity and perseverance. For 250 years, treasure galleons, made of tropical hardwood in the Philippines, shuttled across the Pacific. In the early decades, two ships a year sailed, but this was soon reduced to one ship a year making the 10,000-km (6,000-mile) one-way voyage. The stout-bellied vessel would depart Acapulco in January or February packed with silver, supplies for the Manila colony, foodstuffs from the New World, such as sweet

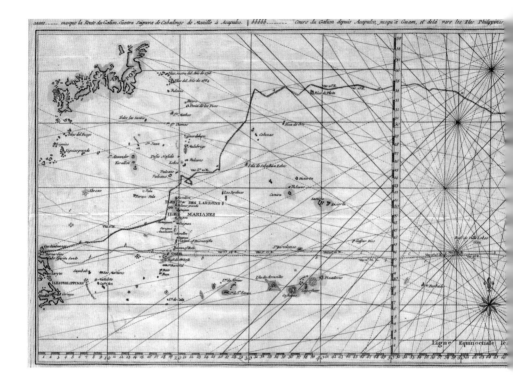

potatoes, peanuts and cassava, and crew and passengers numbering hundreds of souls. Westbound galleons could catch the trade winds and arrive in harbour at Manila in the late spring. Following fevered rounds of negotiation between Fujianese merchants, and agents, buyers and sellers in Manila, the galleon would be repacked with silks, gold, spices, carpets, gems and Malay slaves for its return journey. The precariously overloaded vessel needed to depart Manila by mid-June, in advance of the typhoon season. Some, risking late departure or simply running into bad luck, got caught in the fierce typhoon winds.

Sometimes travellers were forced back to Manila – an economically baneful, last-ditch manoeuvre known as the *arribada*. In other instances they were lost at sea. The Philippine archipelago of 7,100 islands, irregular coastline, coral reefs and submerged rocks was a treacherous tract to navigate. The Manila to Mexico route cut through the typhoon belt; in addition, the

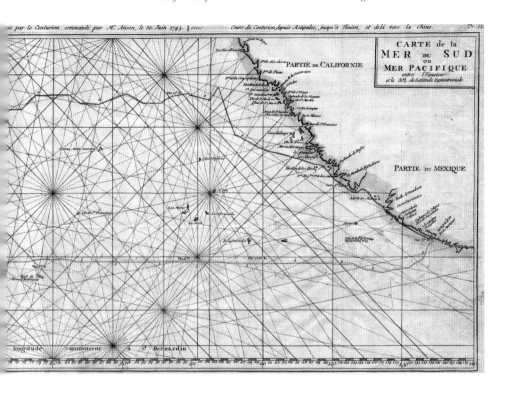

R. W. Seale's map of the Pacific Ocean trade routes from Acapulco to Manila, 1748.

galleons' cargo rendered them large, slow-moving targets for avaricious adventurers. Over the span of two and a half centuries, forty galleons were lost to shipwreck or piracy. The odds of calamity were, then, surprisingly high, and the spectre of financial ruin and loss of life haunted every sailing. Often nothing would have been known until a ship failed to appear either in Manila or Acapulco, when the appalling suspicion of catastrophe would have crystallized with each passing day without news. These treasure-laden galleons were rich pickings for pirates and enemy ships. The English in particular revelled in fanning 'the flame of privateering' and judged that 'The out-ward bound Galeon is far richer than that returning from Minilla [*sic*] and therefore every endeavour should be exerted to get in time to intercept her.'[22] For centuries disruption to the galleon trade was the goal of enemy navigators, and the English managed to capture four galleons. Perhaps most poignant is the story of the unfortunate *Santa Ana*, which encountered a typhoon in 1587 after leaving Manila. After extensive repairs, it was able to resume its voyage, but on approaching the California Baja coast it was attacked by the English navigator Thomas Cavendish and his aptly named ships *Content* and *Desire*. Although the *Santa Ana* was far larger, it had sacrificed cannons for extra cargo space and was no match for the armed English ships. Cavendish loaded as much gold, silk, spices and wines onto his ships as they could carry, and set the *Santa Ana* alight, though he did ferry its crew of 190 Spaniards and Filipinos to a beach. A letter dated 1588 from Manila officials to Philip II grieved, 'this was one of the greatest misfortunes that could happen to this land'; Cavendish returned victorious to Greenwich, the *Desire* outfitted with blue damask sails, and every sailor sporting a gold chain around his neck.[23] The return trip to Acapulco was a long and difficult six months. Because of wind patterns, the galleons were forced to sail far north to the coast of California, and then hug the coast south to Mexico. Typically, some of the crew and passengers fell ill or died en route; on one legendary occasion, a galleon drifted towards Acapulco like a ghost ship, the crew having succumbed to malady while at sea.[24] When the galleons entered Acapulco harbour, they were welcomed by the officials of the royal treasury tasked with ensuring that duties were paid on the cargo. Only after their inspection could the numerous sick be carried to hospital, the healthy begin their procession to the church, and unloading begin. The climax of the homecoming was the Acapulco fair, described by the visiting Prussian geographer Alexander von Humboldt as 'the most renowned fair of the world'.[25] As soon as the galleon was spotted from the northwest coast, thousands of Mexican and Peruvian merchants, indigenous traders, Spanish officials, friars and mendicants began to converge on sleepy Acapulco. On the galleon's arrival, they were joined by Filipinos and Chinese. One account from January 1697 describes the transformation of the 'rustick Village into a populous City' thronged with 'a great concourse of Merchants from Mexico, with abundance of pieces of Eight'.[26] For a few frenetic days, the bounty of the world was exchanged at Acapulco. After the clamour receded, the silks, porcelain, gold and spices were loaded onto mules and carried over the 'China Road' to Mexico City, and beyond to the port of Veracruz, from where they were shipped to Spain. The thirst for Chinese silks in the Americas and Europe seemed as deep as the Chinese thirst for silver. Even after crossing the Pacific and the Atlantic Oceans, Chinese silk commanded a handsome profit. Like silver, silk flowed in rivers, creating cloaks for statues of the Virgin in Potosí, vestments for the priests of Acapulco, gowns for the ladies of Mexico City, stockings for the gentlewomen of Seville and carpets for the wealthy burgers of Amsterdam.

A silver-edged Dutch golden age

Silver flowed into China through Manila, but an estimated 150 tonnes arrived annually via Europe, a significant portion passing through Amsterdam.[27] As the seventeenth century progressed, Amsterdam became one of the world's premier trading centres, attracting an international business class that included ambitious merchants and European bankers. The northern provinces of the Low Countries declared independence from Spanish rule in 1581 (though it would take until 1648 for Spain finally to acknowledge Dutch independence). The rebel noblemen had adopted a silver token in the shape of a half-moon to express their ambitions for their nation. The new Dutch Republic focused its energies on trade and commerce, and rapidly became a capitalist nation, continuing to trounce its enemies through trade.

The founding of the Dutch East India Company (abbreviated to VOC) in 1602 propelled Amsterdam into one of the world's premier commercial hubs, trading spices from Indonesia, porcelain and tea from China, salt and sugar from the Caribbean, fruits and nuts from the Mediterranean, and silver from Europe and Japan to China. Some of these goods were for domestic consumption. Silver graced well-appointed Dutch homes, while framed maps alluded to how such wealth was amassed. But the Republic's strength rested more on its role as middleman,

channelling commodities and luxuries between the Americas, Europe and Asia.

The voc was the world's first vast international corporation, and the lubricator for its business was silver. It created a network of hundreds of trading posts in Asia, the most important being the entrepôt of Batavia (Jakarta), from which it controlled the spice trade. Annually two or three amply armed Dutch fleets made the voyage to Asia around the Cape of Good Hope. A Dutch delegation had established contact with Japan in 1600. They were welcomed positively as trading partners who had no religious interest in the islands, and the Dutch became the only Europeans permitted to trade with Japan. Although confined to Nagasaki, the voc seized and exploited the opportunity to supply China with Japanese silver.

Dutch still-life paintings from the seventeenth century, particularly *pronk*, or depictions of luxury, revel in the bounty

Johannes Vermeer, *Young Woman with a Water Pitcher*, *c.* 1662, oil on canvas.

of the Republic's trade. Tables are heaped with Chinese silks and porcelain, fruits from the Mediterranean, Venetian glass and gleaming silverware – the voc's recrafting of the world reflected, literally, in the virtuoso accomplishments of Dutch silversmiths.

Willem Claesz. Heda, *Still-life with a Gilt Cup*, 1635, oil on panel.

6 The New Flow of Demand

A wagon rumbles over a rutted track, heading westwards with a train of pioneers. It is hot and dusty, and water is scarce, but a jug of milk sits unspoiled under the canvas. At the bottom of the jug is a silver coin. This is familiar pioneer lore.[1] The coin has not been dropped into the milk to hide it from thieves; that would have been as obvious a strategy as slipping a key under a doormat. It is there to keep the milk from turning rancid.

For millennia, silver was valued because it was rare. As convertible wealth, it functioned superbly as coinage, as jewellery and as tableware. Silver was prized for its worth, and secondarily for its alluring, shiny appearance. Today, however, more than half the global demand for silver is for industrial uses – needs that have nothing to do with the metal's rarity or beauty.[2] For decades, the new flow of demand has followed silver's unique chemical and physical properties.

Picture making

Centuries before the discovery that silver salts could miraculously bring to life a photographic image, glass makers were exploiting silver's 'picture making' potential. The Romans had created rudimentary decorative windows by inserting panels of coloured glass into frames, but the art of stained glassmaking reached its apogee in the cathedrals of medieval Europe. Glass is formed by heating sand and wood ash (potash) and allowing it to cool. Medieval craftsmen experimented with mixing powdered metals into the molten mixture, which resulted in glowing blues, reds and browns. The pieces of coloured glass were fitted into lead frames, and details were applied with black paint. But the one colour they had difficulty achieving was yellow – the bright, clear hue essential for a saint's halo, or the rays of the sun.

The solution lay in an entirely different method. Instead of mixing metal into molten glass, a silver compound such as silver nitrate or silver chloride could be added to a clay binder to make a paste. The paste was painted onto clear glass, which was then fired in a kiln. The chemical reaction produced the colour yellow; depending on the temperature of the kiln, the length of the firing and the thickness of the paste, hues from rich yolk yellow to pale luminous gold could be achieved. The earliest examples of silver stain date from the first quarter of the fourteenth century, and craftsmen in France executed work of extraordinary refinement.[3] Instead of piecing small fragments of glass to compose details, the artisan could paint delicate details such as pollen grains on a wild rose, the hair of angels, the gilded edges of book pages or flames licking the feet of unbelievers. As craftsmen gained greater mastery over the technique of silver staining, large quantities of small roundels, made up of single pieces of painted glass, were produced, often for domestic use. From the fifteenth century, silver stain was painted onto blue glass to create details in green, expanding the range of coloured detail without the need for separate leaded pieces.

In the Middle Ages, silver-stained glass could capture an artist's idea of the physical and metaphysical worlds. In the nineteenth century, silver would capture reality itself and crystallize the past in a photograph. Photography developed out of a desire to reproduce external appearance, and was carried forward by a combination of contemporary interests in art and science. On the one hand, artists had long experimented with optical devices such as the camera obscura to assist them in depicting the exterior world; Dutch interiors and streetscapes of the seventeenth century have been scrutinized as examples of this use. Later on, in 1833 the amateur artist William Henry Fox Talbot expressed his frustration with the results of the camera lucida,

The Crowned Virgin and Child as 'The Apocalyptic Woman Clothed in the Sun', Germany, late 15th century, colourless glass with vitreous paint and silver stain.

which projected an image onto a sheet of paper for the artist to trace. Regarding his dismal efforts to convey the splendour of the scenery around Lake Como, Italy, he concluded, 'How charming it would be if it were possible to cause these natural images to imprint themselves durably, and remain fixed upon the paper.'[4]

In the scientific world, chemists had observed that when silver salts were exposed to light, they turned black. Artists had optical devices that could capture images, but no means to save them; chemistry seemed to proffer a solution. Early experiments projecting pictures onto surfaces treated with silver salts were tantalizingly close. Ghostly images appeared momentarily, but then the entire surface would blacken in light. A breakthrough was made in France by the theatrical designer Louis-Jacques-Mandé Daguerre, whose flair for visual drama alerted him to the possibilities of the medium. Through a partnership with the amateur scientist Nicéphore Niépce, he devised a method of developing an image on a copper plate coated with silver salts, and fixed it with a salt solution.

In England, Daguerre's success galvanized Talbot, who had been working for several years on a method of coating writing paper with salt and silver nitrate, and placing the sensitized paper in miniature cameras (or 'mousetraps' as his wife called them). The final piece in the puzzle was the use of hyposulphite of soda as a fixing agent, suggested by Talbot's friend, the astronomer Sir John Herschel. 'Hypo' was immediately adopted as the most effective way to stabilize a photograph and remained so for decades. The triumph of photography, then, was a marriage of chemistry and art.

From the mid-nineteenth century, the most common printing process was the albumen print, in which paper was prepared by coating it in an emulsion of egg white (albumen) and salt, followed by an immersion in a bath of silver nitrate. Although albumen prints were very prone to fading, they had their own beauty, with creamy highlights and velvety brown shadows. By the end of the nineteenth century, gelatin silver prints superseded albumen prints and remained the dominant process for decades. Gelatin, an animal protein, was used as the binder for silver salts, and the resulting prints were capable of portraying a wide range of hues, from rich black to pristine white. Thanks to silver, then, we know what the French painter Rosa Bonheur looked like, and we can see the now-lost landscapes of major world cities: Paris before Haussmann, San Francisco immediately after being flattened by the 1906 earthquake and Shanghai before its rapid modernization. We know how our great-grandparents looked as children, and can reminisce over how our own children looked in their infancy, or how we ourselves were in our younger years.

Since the invention of the daguerreotype in 1839, processing the world's photographs has consumed a significant amount of silver – over 6,000 tonnes a year at the start of the twenty-first century. Silver is a component of photographic film and paper, as well as of plates for X-rays. A decade later, this figure was slashed by two-thirds, with the rising popularity of digital photography.[5] Most photographs taken today are never printed,

Julia Margaret Cameron, *Child's Head, Freshwater*, 1866, albumen silver print.

though they may be shared by hundreds of thousands over social media sites. As will be discussed later, the electronics involved in the taking and viewing of digital photographs may consume silver, but only at a fractional amount of that which sluiced through the developing pans of professional and amateur photographers for nearly two centuries.

Walker Evans, *Young Girl*, c. 1936, gelatin silver print.

Healing

Silver preserves our memories, and also our health. In a collection of the Barnes Foundation in Philadelphia is one of the most familiar paintings of the modern period: Matisse's *Le Bonheur de vivre* of 1906. It is an artwork that celebrates humanity in its prime, infused with health, vigour and vitality. The clear skin of its joyful subjects, who dance, play music, embrace and recline, is all the more glowing in the context of the Edenic landscape's clear, pure tones of yellow, orange and green. This is apposite, given that the painting's existence in this collection is due ultimately to the health industry, and especially to one silver-based product that made its creator the fortune which funded his collection.

In the 1880s a German obstetrician, Dr Carl Crede, introduced the use of silver nitrate eye drops to prevent eye infections contracted when newborn babies were exposed to bacteria in the birth canal. The results were impressive; the incidence of infection, which could lead to blindness, rapidly declined from nearly 8 per cent to just over one in a thousand.[6] The downside, though, was that silver nitrate was an irritant to delicate tissue, which could itself lead to infection. Aware of the scale of the potential market, pharmaceutical companies scrambled to create a safer product. In 1902 the American doctor Albert Barnes and his German colleague Hermann Hille formulated a compound of silver and wheat protein they named Argyrol, which they heavily promoted to influential surgeons and physicians. An astute entrepreneur, Barnes grasped the opportunity to commercially exploit Argyrol on a global scale, and formed a company with headquarters on three continents. Although it had competitors, Argyrol quickly emerged as the market leader in some regions, especially when laws were enacted requiring newborns to be treated with antimicrobial eye drops. By 1907 Barnes had become a millionaire.[7]

Still a young man, he decided to use the profits from the sale of Argyrol to build an art collection. Employing his usual business acumen, he focused on the undervalued and underappreciated work of the Impressionists and Post-Impressionists, amassing a collection both priceless and unreproducible, which has long been the envy of museums worldwide. That this collection exists at all is because one of our grandparents' first confused sensations, as they emerged into the world, was the nip of a silver compound in their eyes.

Of course, neither Barnes nor his nineteenth-century predecessors were the first to discover silver's antimicrobial properties.

Child's medicine spoon, 18th–19th century, Europe, silver.

For thousands of years people had observed that silver could preserve water, milk and wine. Early physicians wrote of its efficacy in staunching infection and disease, and used silver implements in their fight against disease. Hippocrates, the father of Western medicine, advised sprinkling powdered silver on ulcers.[8] Historic pharmacopoeia listed all sorts of inventive uses for silver, including treatment for burns, blood disorders, epilepsy, syphilis, heart disease, halitosis and even the Black Death. Recognizing that silver didn't usually cause allergic reactions, early surgeons experimented with silver prostheses. These might have appeared incongruous – the influential Danish astronomer Tycho Brahe became known as 'the man with the silver nose' after losing part of his nose in a brawl and replacing it with an electrum (silver/gold) prosthesis. When it was destroyed one night by his pet dog, he had fourteen more manufactured, one of which wound up in the possession of Voltaire.[9]

Some of these applications enjoyed a level of success; many did not. A considerable stride forward came when a surgeon from Alabama devised a successful means of suturing wounds with silver wire. J. Marion Sims, now considered the founder of surgical gynaecology, practised in the slaveholding society of the mid-nineteenth-century American south. Moved by the plight of female slaves who suffered catastrophic internal injuries during childbirth, he created a surgical method to reverse the incontinence caused by prolonged obstructed labour.[10] Initially, his attempts with silk sutures were unsuccessful and always resulted in infection. Eventually, he turned to a jeweller who crafted for him fine pure silver wire to use as thread. At last, in 1849, four years after his first attempt with silk, he cured a young woman who had endured 29 unsuccessful surgeries.[11] It was a triumph for the young surgeon, who went on to found the Woman's Hospital in New York, and to demonstrate his technique with silver wire sutures throughout Europe.

All this Sims accomplished without an accurate grasp of germ theory – that would have to wait several decades until the pioneering laboratory work of Louis Pasteur and Robert Koch uncovered the pathogens that cause disease. Following their breakthrough, in the first decades of the twentieth century, silver foil was being used to dress wounds and inhibit infection, and silver nitrate was an accepted therapy for burn injuries and skin conditions. It was also an effective treatment for many eye disorders. During the First World War, silver compounds were routinely applied to wounds to cleanse them. Colloidal silver (ultra-fine silver particles suspended in liquid) was widely consumed as a panacea for a cornucopia of ailments. It is still a heavily promoted, though controversial, health supplement today.

The availability of antibiotics from the 1940s resulted in decreased use of silver in healthcare, but since the 1990s the

A silver-plated artificial nose made between the 17th and 19th centuries.

metal has reappeared in coatings for medical devices, bone cements and bandages, and on hospital surfaces, water filters, bedrails, door handles and clothing. One impetus behind its increasingly ubiquitous use in the fight against bacteria has been the development of nanotechnology, which is the manipulation of matter at a vanishingly small scale, generally under a hundred nanometres. To appreciate just how small this is, one nanometre – a billionth of a metre – is how long a fingernail grows in a second. By way of comparison, a human hair is about 100,000 nanometres in diameter.[12] Although nanotechnology is new, nanoscale particles are not; scientists had used them for over a hundred years without understanding their make-up. What is new is the way these tiny particles are manipulated and synthesized. Chemists and materials scientists have found new ways to bundle small packages of atoms and incorporate them into plastics, textiles, coatings and gels. At nanoscale, silver's toxicity to bacteria is even more pronounced, and the result has been an explosion of 'antimicrobial' products whose range has expanded far beyond traditional healthcare. Nanosilver now appears in hundreds of consumer products, from athletic clothing – in which it helps by inhibiting the bacteria responsible for body odour – to yoga mats, underwear, swimming pool filters, bed linen, cosmetics and baby bottles. Nanosilver is in bus handrails, Korean toothpaste and disinfectant spray on the Hong Kong subway.[13] Ceramic water filters coated in nanosilver have been promoted as a silver bullet for the problem of supplying clean drinking water to the developing world. But our faith in its capacity to gobble microbes has resulted in silver ending up in places it isn't usually found – including our bodies.

Most of us have low levels of silver in our blood or tissues. We absorbed it through our antifungal sports socks, swallowed it with our tap water or inhaled it on public transport. Low levels of silver exposure do not appear to cause us harm. Absorb too much, though, and the result is argyria, a non-fatal though cosmetically embarrassing affliction which turns the skin blue. Silver's impact on the environment, however, is another story. Under certain conditions, an accumulation of silver in sediments and water can pollute water sources and pose a serious risk to aquatic life. For example, the almost-universal popularity of traditional photography during the 1970s and 1980s, and lax environmental oversight in some countries, resulted in silver contaminating waterways and degrading the environment.[14]

While we know a certain amount about the effect of silver on the human body and the environment, our knowledge of the impact of nanosilver lags far behind. Our experience with the overuse of antibiotics reminds us that we can have too much of a good thing; there are mounting concerns that frivolous use of nanosilver might result in a new breed of silver-resistant bacteria.[15] While we still know too little to make an accurate risk assessment, the appearance of nanosilver products proliferates, and the flow of demand carries silver down drains, along rivers and through our bloodstreams.

Flow

It's a familiar scene throughout the world: the congested freeways of Los Angeles, the bumper-to-bumper highways of Mumbai, the stationary elevated expressways of Shanghai, the choked urban highways of Mexico City and the clogged London ring roads – traffic jams define our cities. But while traffic may be caught in gridlock, the silver inside each car ensures the smooth, unimpeded functioning of its electrical components. The Rolls Royce 'Silver Ghost' might be the most prestigious car in automobile history, but even the most ordinary car today could contain 16 g (0.5 oz) of actual silver.[16] Up to forty silver-tipped switches help us to start our cars, adjust our power seats, turn on our lights, open our power windows and defrost our rear windshields. Silver is a superb conductor of electricity, and the electrical and electronics industries made up about 45 per cent of global industrial demand in 2014.[17]

We rely on silver not only on the ground but in the air. Aeroplane engines, running continuously at high temperature, rely on silver-coated bearings to run safely. As silver resists corrosion and creates low friction, it can act as a lubricant between fast-moving steel parts. The amount is tiny – a few nanograms per bearing – but the impact on the smooth running of heavy machinery under high-stress conditions is enormous.

Many of the taken-for-granted conveniences of contemporary life rely on silver. When we tap our mobile phones to switch them on, or touch the controls of a microwave oven, television or thermostat, we communicate through a silver membrane switch. Because of silver, we can check our e-mail while heating up a meal in the microwave, and adjusting the temperature in the kitchen. Our most essential consumer items – laptops, desktops, phones, televisions – use circuit boards printed with silver ink which create the electrical pathways fundamental to our interconnected world. Technology might well come up with an alternate solution, but until then, we are tethered to the white metal.

When silver is used in a switch- or circuit board, its structural and chemical properties ensure the uninterrupted flow of electricity; as a lubricant, it keeps machines in motion. In the form of silver-oxide button-cell batteries, it keeps our watches in motion, and calculators powered. But it has also been used not just to facilitate the flow of energy but to capture it in the form of solar energy. Currently, the most widely used solar panels contain photovoltaic cells in which silver paste is used as a conductor for an electric current. It has been estimated that 2 billion grams (70 million oz) of silver are projected to be used in solar panels by 2016.[18]

In the past, silver in the form of currency lubricated the flow of ideas, ideologies, armies and trade. Today silver-printed circuit boards fulfil much the same function. Ideas of democracy, entrepreneurship, terrorism and philanthropy can travel anywhere, through the agency of silver.

7 Status Symbols

We coin currency from silver, but we also coin idioms: 'born with a silver spoon in one's mouth', 'presented on a silver platter', 'every cloud has a silver lining'. From silver-tongued to silver bullet, our language is rich in sayings that point to the complexity of meanings we give to silver.

Meanings are, of course, made. To be able to speak of 'the family silver' rather than, say, 'the family tin' requires a long human history of mutual agreement and discernment. We can say that silver and gold are symbols of status and power, but more interesting is the question, 'Why?' And how have they continued to hold fairly stable meanings through millennia and across very different cultures? Two ideas about silver that seem as ingrained as the fact that it shines are the notion of status, and the concept of purity.

Silver has been used as a marker of status ever since humans learned how to extract it from the ground. As silver-bearing ores are localized and not distributed uniformly over the earth, silver was recognized early on as rare and precious, and something that was refined with effort and ingenuity. The clean, even gleam of a lovingly polished surface made a silver object all the more covetable. Possessions confer status on their owners because we transfer value from the object to the person. We prize rarity. But we also prize things that shine, and silver is, to human sight at least, the shiniest of metals. Recent research in evolutionary psychology suggests an intriguing reason behind our attraction to sheen: gleaming surfaces remind us of life-sustaining water.

Moreover, just as control of clean water has been a mark of power, the ownership of shiny, desirable objects is deeply embedded in our brains as a mark of prestige.[1] For all that we equate glitter with sophistication, it may well be our lizard brain that drives us.

Leslie Lewis Sigler, *Family of Heirlooms*, 2011, oil on canvas.

Status in death

Silver and gold have been status symbols for millennia, and for much of this time it has been firmly believed that these metals would be precious throughout eternity. Beliefs and practices are of course myriad, but common threads persist: the practice of burying the rich and powerful with luxuries and precious metals was an acknowledgement of one's high rank in life, and an assurance of continued good standing in the afterlife. The ability to attend the everlasting feast sumptuously attired and with one's own good plate has been a fairly constant preoccupation of the living in their quest to prepare for what might follow.

At the ancient Mesopotamian site of Ur (now in southern Iraq), sophisticated silver and gold objects from around 2500 BCE were excavated by Leonard Woolley in the 1920s and '30s – the 'Golden Age' of handsomely funded archaeological digs. (A famous visitor to this site was Agatha Christie, who married Woolley's archaeological assistant.) Possibly the silver had been traded from neighbouring Anatolia, rich in silver-bearing ores. In ancient Mesopotamian belief, the universe was a sphere, divided between the world of the living and that of the dead. The realm of the dead was particularly cheerless, the food unpalatable and the water bitter. The fluted silver cups, jewellery and silver boxes for eye kohl found in the burial pits may all have been essential creature comforts provided in anticipation of grim conditions. They may also have been gifts for underworld deities; music from wooden lyres covered in sheet silver and embellished with shells and lapis lazuli might well have been expected to quell the tempers of irritable gods provoked by the harsh environment.

Where cultures considered boundaries between this life and the next to be porous, understandable emphasis was placed on securing safe and comfortable passage. The ancient Egyptian preoccupation with death might conjure images of Tutankhamun's iconic gold mask, but in earlier periods of Egyptian history, silver was more highly valued, partly because it was rarer. The skin of the gods was gold, but their bones – perhaps the more fundamental part of the body – were silver.

Lacking substantial local silver deposits, the ancient Egyptians might have imported the metal from Greece, but they could also have found it in their own gold mines. In nature, silver and gold form a series of alloys. These range from 'aurian silver', defined as silver containing over 5 per cent gold, to argentian gold, with about 20 per cent silver – more commonly known as electrum. Analysis of silver objects excavated from Egyptian tombs seems to support the argument that ancient Egypt's own silver-rich gold ore supplied the country with the silver used to fashion necklaces, scarab amulets and coffin fittings appropriate for persons of high rank.[2] Tellingly, the Egyptians' earliest term for silver was 'white gold'. One of the most dazzling collections of silver ever found in a grave was that discovered at Sutton Hoo, East Anglia, England, in 1939. A sumptuous burial chamber in a 27-m (89-ft) boat (a ghostly imprint is all that now remains of the wooden vessel) was the final resting place and treasure trove of a man who might have been an East Anglian king. Among the gold, gems and rich textiles was a hoard of finely crafted

Statuette of
a woman, Egypt,
610–595 BCE, silver.

silver vessels that were likely gifts from the Byzantine Empire.
The epic Anglo-Saxon poem *Beowulf* brings to life a convivial
context for these riches. These are the silver bowls, cups, spoons
and silver-mounted drinking horns that belong in the great mead
hall. While the Sutton Hoo burial chamber is a hoard of private
wealth, it is also a symbol of a generous and heroic host readying
the banqueting hall to reward loyal followers.

Perhaps the most delightfully envisioned afterlife was that
crafted by the artisans of China's Tang Dynasty (618–907 CE).
Following earlier traditions and beliefs in the afterlife as an
extension of the present, Tang nobility outfitted themselves with
luxuriously appointed multi-chambered tombs filled with treasures and daily necessities. Tomb furnishings ranged from the
pragmatic – such as clay horses and chariots for transport, multi-
storey villas in miniature, food and wine – to luxuries such as clay
figurines of foreign musicians recruited from Silk Road cities,
fine silver and gold tableware, jewellery and cosmetics. During
the Tang Dynasty, the Chinese Empire was the most extensive
of the medieval world, and trading networks extended into
Persia, Japan, Korea and Vietnam. The capital at Chang'an (now
Xi'an) was a cosmopolitan metropolis and its elite were buried
with artefacts whose origins and ornament reflected the vast
range of the empire. For instance, tombs excavated around Xi'an
turn up silver ingots; Japanese coins; Persian silver coins; Byzantine gold; Persian silver plates, cups and bowls; boxes for
toiletries and delicately crafted silver jewellery.[3] In Tang-Dynasty
China, everyday essentials were afterlife essentials.

To have such ample wealth that quantities of treasure
could be buried with the dead might have inspired awe, but it
also attracted censure. There were frequent official injunctions
against the long processions of grave goods that were carried so
publicly from the capital out to the tombs, which underscores
the notion that the celebration of status depends – largely – on
display, and viewers primed to draw comparisons.

Table manners

Silver scissors, China, Tang Dynasty, 618–907 CE.

Often the silver items found in tombs have been associated with dining. The idea of the eternal banquet is shared across many religions; the centrality of sustenance to our earthly existence spills into the future. But *dining* meets other needs that are fundamental to our survival, such as belonging to a group, the ability to nurture others and, on a more sophisticated level, the opportunity for conviviality and the chance to define ourselves, to offer *largesse* as host – to impress or to compete. For this we need the commonly agreed-upon markers of status such as costly and difficult-to-procure foodstuffs, elegantly dressed attendants and lavish plates.

As we have seen, silver's rarity and costliness would always define its prestige. Further, its malleability and lustre would mean it could be worked into forms that reflected the tastes, learning and fashion-consciousness of its owners. And what better arena to display silver than at the table, where the public machinations of status are inescapable? Where one sits, next to whom, how close to the host or to objects as seemingly innocent as the salt cellar, what food is served, on what the food is served and who is served first – all conspire in the theatre of status that is a dinner party.

Though fictional, the aspirations of Emma Bovary illuminate the seductions of silver at the table. Gustave Flaubert's *Madame Bovary* (1856) may be the consummate novel of sex and shopping, but thrown into this heady mix is silver. Soon after her marriage to a plodding provincial doctor, Emma realizes her romantic aspirations for wealth and status remain far beyond her grasp. Her problem, as we might say today, is thwarted self-actualization. A rare invitation to a ball leads her to a dinner table laden with an intoxicating mix of delicate meats, truffles, lobster, quail, fragrant bouquets, fine linens and silver objects reflecting the candlelight

A lavishly silver-laden table. Table service, commissioned by Friedrich Wilhelm von Westphalen, bishop of Hildesheim, Germany, 1763, silver.

– shaping and reflecting an image of the lavishly romanticized existence for which she yearns.

Galvanized by this experience, she falls back on the most common of ploys – advertising her status through purchases she cannot afford. When she takes her first lover, it is silver (a riding whip with a silver-gilt handle) that she gives to him as a love-token, and to symbolize her own refinement. Silver as a status symbol is strewn along the route of Madame Bovary's infamous downfall. It gleams in jewellers' windows and on dining tables more sumptuous than her own; she pawns silver wedding presents to stave off debt. She envies the scrupulously polished silver on the table of the village notary – even as she implores him to save her from ruin.

The Roman table

Imitation is a simple (and as Madame Bovary discovered, sometimes dangerous) form of flattery; the easiest way to emulate those we admire is through possessions. It is an impulse that cascades down the centuries with greater or lesser success, but it is also a compulsion that has given us some of the most sophisticated silver of the ancient world. The Roman elite, and

Pair of cups, Roman, late 1st century BCE–early 1st century CE, silver with gilding.

their emulators, heaped their tables with silver. Several archaeological discoveries during the past two centuries have unearthed treasure troves of late Roman banquet silver unparalleled in its heft and exquisitely wrought ornament. The Mildenhall Treasure now in the British Museum, discovered during a felicitous ploughing session in 1942, consists of an impressive cache of silver platters, plates, bowls, dishes, ladles and spoons. The iconic 'great dish' measures an impressive 60 cm (24 in.) in diameter, and portrays the riotous merrymaking of Bacchus, a drunken Hercules, Pan and swirling maidens. Bacchus is a Roman god, but his origin was the Greek Dionysus, just as Hercules was the Greek Heracles, and those lustful satyrs descended, with Pan, from Greece. By this stage, the Greek world and its elite dining rituals fascinated the Romans, not just in Rome but throughout the empire. On art objects, wall paintings and tomb art, images of reclining banqueters being served food and wine from objects very much like those found at Mildenhall recalled a 'Golden Age' of privilege and indulgence. The owner of the 'great dish' possessed not just wealth but membership of a class that shared admiration of ancient Greece (or, more accurately, how it was imagined).

This enormous platter has a counterpart in an even larger dish (70 cm, or 28 in., wide) that is part of the Sevso Treasure, from the same period. The Sevso 'hunting plate' has an outer frieze depicting hunters on horseback and on foot pursuing prey through the woods. A central medallion portrays picnickers enjoying an al fresco feast under an awning, attended by servants and overlooking the boar, deer, fish and game that stock their estate. Encircling this scene of bucolic bliss is an inscription:

> May these, O Sevso, yours for many ages be
> Small vessels fit to serve your offspring worthily.[4]

We do not know the identity of Sevso, or why his 'family silver' was buried in Eastern Europe, where it was recovered in the 1970s under circumstances so murky they resemble a fictional international art theft thriller.[5] We know nothing of his life except the image that he chose to project for posterity. Just like Emma Bovary, Sevso found in silver a way to imagine a more abundant life.

What were such extravagant vessels used for? Likely they were prominently displayed on special stands, as we can see on surviving paintings. This in itself harked back to Greek and Etruscan customs. They were almost certainly used at banquets for opulent effect. At a seriously sybaritic feast, silver platters might have been laden with roasted thrush, geese force-fed with figs (an ancient foie gras), stuffed dormice, piglets or ducks in prune sauce.[6] Spices, such as pepper from India, might have been served in whimsical pepper pots. Wine would have been mixed to each drinker's taste with warm or cold water, and imbibed from silver cups.

All this banqueting silver was used for visual effect, for the transport of food and for amusement. Roman banqueting silver is laden with all sorts of clever references to deities and demi-gods that would have been intended to spark lively discussion between peers. Education, shared culture and mutual delight in classical references all contributed to making these silver objects symbols of 'soft power' that was just as enviable as buying power.

Sometimes this silver had a religious purpose too. In 1830 at Berthouville, Normandy, a farmer in another lucky ploughing episode discovered a cache of silver bowls, cups and jugs, now considered one of the most remarkable troves of Roman silver yet discovered. This region was under Roman control by the first century BCE, and the silver was dedicated to the local version of the Roman god Mercury. Like many silver items found at the site of Roman shrines, much of this silver was originally domestic, and decorated with the familiar classical banqueting themes. Some of these pieces are inscribed with the names of donors, whose names tell us they included Roman citizens as well as locals. We might imagine the owner of a silver wine cup, de-ciding out of piety to donate an exquisite favourite possession to the sanctuary. But given the prominence of the inscriptions, some of which were inlaid in gold, the wish to display status probably played a role. The inhabitants of provincial Berthouville were no less eager than those closer to power centres to keep their names in the public eye at a sanctuary used for pilgrimage and feasts.

The French table

The Roman feast might now be a byword for gluttony and riotous excess (perhaps justifiably – the stoic statesman Seneca was scathing about slaves being engaged just to clean up vomit). But it was also an art. Perhaps the culture that best compares in an understanding of dining as theatre is eighteenth-century France – another society condemned for inexcusable self-indulgence.

Before the Revolution, the tables of the French aristocracy had glittered with silver, and Parisian silversmiths were sought out by European courts. As banquets became ever more extravagant, matching silver dinner services became a requisite accompaniment to sophisticated cuisine. Most eighteenth-century services included up to 250 pieces; one ordered by Catherine the Great from the workshop of the Paris-based Flemish silversmith Jacques-Nicolas Roettiers was comprised of over 3,000 items, to serve sixty diners.[7] It included tureens, plates, sauceboats, spice boxes and 24 uncannily realistic scallop-shell dishes, with handles resembling seaweed, and whelks clamped to the shells' ridges. She presented it (in gratitude?) to her lover Grigory Orlov, who had been involved in the coup d'état that led to the abdication and assassination of her husband, Peter III, and her own accession as ruler of the Russian Empire.

Not much French eighteenth-century silver survives; it was either melted for bullion to finance military campaigns or destroyed during the Revolution. Rare pieces made for foreign clients and inventories probably give us the best idea of the extravagant use of silver at the eighteenth-century French court. One inventory of a nobleman's private silver collection at Versailles even lists a silver bidet, whose fate, sadly, remains unknown.[8] The dining habits of the French aristocracy influenced customs throughout Europe. The type of service itself – the *service à la française* – necessitated a panoply of specialized serving vessels and dishes. For this type of service, which

persisted into the early nineteenth century, meals were served in a prescribed set of courses. For each course, all the dishes were brought to the table together and arranged decoratively. The silver and the cuisine, like the diners, were on stage. And the practice of having the family coat of arms prominently engraved on the silver reminded guests and host exactly whose stage they occupied. As each guest helped himself, and dishes were not removed from their positions, multiple offerings of each dish were placed on the table, along with multiple accessories such as sauceboats and servers for condiments. The tables of the nobility groaned not just with food, but with silver.

Distinguished guests might be seated by sculptural centre-pieces. As soup began every meal, the tureen was an essential vessel. This massive object afforded the celebrity silversmith an opportunity to display his virtuosity. Intended to hold food such as shellfish soups or rich game stews, the most lavish tureens were sculptural extravaganzas depicting fowl, fish and crustaceans. The still-lifes on the lids, meant to announce the contents inside, were sometimes cast from life, and then worked with hammers and other tools to reach a level of realism that

Jacques and Jacques-Nicolas Roettiers, soup tureen on a stand and four candlesticks from the Orlov service, 1771–3, silver.

today we might find more disconcerting than appetizing. Who now wants to be reminded of the dogs that bring down the stag or the rabbit waiting, limp, to be skinned?

One menu for a relatively modest dinner party for twelve people, suggested by a mid-eighteenth-century cookbook, reveals why these meals and the requisite service placed such demands on plate.[9] The first course includes a herb soup, a rice soup, hors d'oeuvres and a joint of beef. In addition to tureens and a platter for the beef, individual saltcellars, sauceboats and mustard pots likely accompanied the display. Mustard could be powdered and sprinkled over roasts with a silver castor, or ground into a pungent paste with wine or vinegar and served from a pot. Hors d'oeuvres, especially if they contained shellfish, might have been served on silver scallop dishes. These scallop dishes are often renamed 'butter dishes' by the antiques trade to improve their saleability, but were never originally used for that purpose.

The second course was comprised of truffled, braised veal tenderloin, mutton chops with basil, ducks and a braised hen, all requiring platters and serving tools. These would all be replaced

Jacques-Nicolas Roettiers, pair of scallop-shell dishes, 1771–3, silver.

for the third course of a roasted young hare, roasted pigeons, vegetables and a chilled custard. The dessert course included fresh fruit, two dishes of stewed fruits and plates of waffles, chestnuts, gooseberry jelly and apricot marmalade. These might have been displayed on an epergne – a theatrical centrepiece, which appeared first at Versailles, for the offering of delicate hothouse fruits, jellies, nuts and confectioneries. All in all, such a dinner was a gastronomic cornucopia requiring a pantry of plate.

Keeping food and drink at the optimum temperature resulted in new types of dish, such as the monteith – a large bowl filled with crushed ice for chilling wine glasses. The glass stems were supported by the monteith's notched rim. The name was supposedly derived from that of a Scotsman, Mr Monteith, a fashion maverick who favoured coats with scalloped lower hems – reminiscent of the rim of the bowl. The owners of such *au courant* objects might have gazed with satisfaction at their reflections in the polished surfaces – and possibly at the envy reflected in the reactions of their guests.

Edme-Pierre Balzac, tureen, 1757–9. The ornamentation shows a stag being dragged down by hounds.

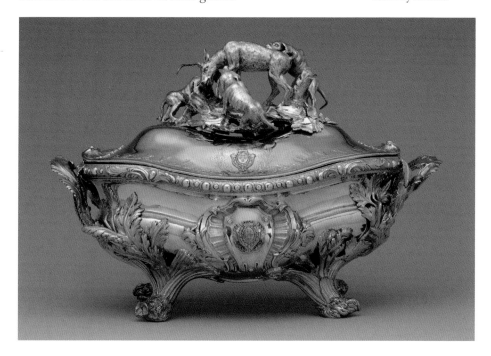

When we see these silver items in museums, or reproduced against a photographer's white backdrop, it is tempting to view them simply as artworks, which they were, but just as importantly they were part of the drama and choreography of the party. It is easy to forget that the platter engraved with a coat of arms was once piled 'with a quivering hindquarter of fat beef encircled by a simple vegetable garland studded with lardoons';[10] that the pristine silver spoon was once heaped with stuffing made from parsley, butter and the chopped organs of woodcocks; or that the smooth, polished belly of a bowl was beaded with the evaporated steam from a tureen of crayfish soup, whose lid had just been lifted. The rococo scrolls of a pierced silver basket were not curved lines in space, but cradled the soft pink skins of ripe peaches. And congealed butter dripped from the lips of sauceboats. All this silver would glisten under the light of candles supported in handsome silver candlesticks. We almost never see such a display now the way it was routinely enjoyed – that is, by candlelight. Under these conditions, the scrolls, flowers and animals so skilfully conjured by the silversmith would have been animated; warm, directional candlelight would have enhanced the sculpture and drama in a way we miss when we see the pieces under even, electric light; even the underbellies of tall vessels would have been alive with detail. Without candlelight, we lose the movement and crevices of darkness out of which illuminated surfaces gleam.

Good standing in Britain and America

The increasing popularity of new foodstuffs led to the refinement of new types of silverware. In Britain, a whole genre grew up around the drinking of tea. While coffee was more commonly imbibed in public coffeehouses by men, the ritual of tea-drinking was cultivated by women in their private rooms, as well as men. Tea caddies held the expensive loose-leaf teas imported from Asia, and large kettles on stands provided a steady supply of hot water. When tea was at its most expensive, rebrewing (as a form of rationing) made for an insipid beverage. The shapes of teapots,

sometimes with fruitwood handles, morphed with changing fashions. At certain times in the eighteenth century, the fashion for adding warmed milk to tea required covered hot milk jugs. Spoons required spoon trays. And the irritating tendency of tea leaves to migrate to where they were unwanted required a mote spoon with a long, pointed stem to unclog the pot's spout, and a pierced bowl to skim leaves and tea dust from the drink's surface.

Workshop of John Leach, monteith bowl, 1703–4, silver.

This specialization of type became a fetish during the Victorian era, when a misstep could spell ruin for the status-conscious hostess. To help her, etiquette books and magazines offered advice on the proper knives for fish, or the indispensability of a double-ended oyster fork with a pierced bowl to strain liquid and grit on one end, and tines to impale the flesh on the other. By this time, the *service à la française* had given way to the *service à la russe*. Now food was carved by servants at the sideboard, and brought plated to each diner. Instead of platters of food occupying the table, there was now more room for silver.

By the second half of the nineteenth century, the dining room had become one of the most important spaces in the middle-class house; what and how to eat were absorbing concerns. While our current culinary obsessions reveal themselves in food blogs and chef idolatry, the nineteenth century had dining paraphernalia – preferably in silver. 'Dinner Giving', one etiquette manual for the aspirational classes reminded its readers, was 'not only a test of the position occupied in society by the dinner-giver, but . . . a *direct* road to the obtaining [of] a footing in society' – a test indeed.[11] Both host and guest could just as

Hester Bateman, tea
caddy, 1788, silver.

easily be wrong-footed by ignorance of cutlery convention, such as knowledge of the angle at which the fork should approach the mouth, or failing to use a fork to eat blancmange, jellies or iced puddings and erroneously opting for the dessert spoon. The knife, obviously, ought never to approach the lips (to contrast with the regressive Victorian, attempting to lift peas to the mouth, with the back of the blade). But when was one to use each of the three knives that convention dictated should be laid in a set? It was permissible to cut sweetbreads with a knife, but rissoles were only to be eaten with a fork. Tips of asparagus ought to be cut with a knife, and to eat an asparagus spear with one's fingers was an unforgiveable faux pas – good manners proscribed the touching of food, hence the baffling artillery of cutlery and tools. Ever more specialized forms evolved to confound the uninitiated. A pair of grape shears, for example, was not simply a utilitarian device decorated with vines as a helpful clue to its function, but a cutting tool to divide the untutored from those more au fait with proper manners. After the stems had been cut to appropriate length, the diner was advised to keep a half-closed hand by her mouth, to more easily convey grape skin and seeds back to her plate without detection by fellow diners. The paintings of the contemporary American artist Leslie Lewis Sigler, which ruminate on specialized silver implements whose purpose has long been for-gotten, poke gentle fun at all this vexation.

The social mobility of the age exacerbated status anxiety, and seemed to inject an unhealthy regard for the possessions of others. It was not the upper classes who worried about how to distinguish a fish knife, such items being considered bourgeois. Etiquette manuals were written for those on their way up, and if the correct use of silver berry spoons, bouillon spoons and bonbon spoons could solidify one's 'footing', then all the more

Mote spoon, British, 1760–80, silver, used to strain tea leaves.

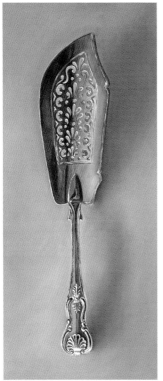

Leslie Lewis Sigler,
Three Graces, 2014,
oil on panel.

reason to prepare the next generation as early as possible. It is not accidental that child-rearing manuals were an equally popular publishing niche, recommending that children be acclimated to the dinner table as soon as they could hold a spoon.

Popular items for infants were silver porringers and baby spoons. These silver spoons were small miracles of engineering, with handles looping back almost to the bowl to accommodate tiny fingers. The bowls were deep and perfectly shaped to discourage spills. The design began to appear in America in the 1890s, and was accompanied by an array of silver specially designed to help infants master table skills. Maybe the parents of these infants also intuited a modern child development theory: possessions help infants develop a sense of self.

Giving silver as a gift to a baby is a tradition in Europe that reaches back at least to the Middle Ages. Often the gift, from godparents, marked a well-born child's baptism and took the form of a silver spoon – a practice that probably underlies the idiom 'born with a silver spoon in one's mouth'. Much of the silver given to babies towards the end of the nineteenth century had the specific aim of encouraging good manners. Both the way it was advertised and child-rearing manuals stressed the 'training' function of children's silver. The silver 'food pusher' could be used to shovel food towards a spoon. Some spoons had animal motifs raised along one edge – cunningly thwarting ill-mannered left-handedness: the barrier along the right edge of the bowl made it impossible for a 'lefty' to tip food into his mouth.[12]

During the Victorian era, the mass manufacture of children's silver began in earnest. The Gorham Manufacturing Company was founded in Providence, Rhode Island in 1831, and became one of America's largest manufacturers of silverware, supplying such prestigious households as the White House. The catalogues produced by the company towards the end of the nineteenth century reveal the multitude of tableware and children's gifts now marketed to the public. Silver – even if it were just an item of tableware or a baby gift – was increasingly within reach, and the reasons for this lay in the amount of silver on the market after the opening of the enormously productive mines in the American West, and changes in the way silver goods were manufactured and distributed.

The metalworking trades were industrialized relatively early. There were of course high-profile silversmiths (in Britain, some of these even became wealthy landowners) but they led sometimes very large workshops. Silver-working was a business producing goods not just for a wealthy elite, but increasingly for the popular market. Throughout the eighteenth century, division of labour became standard. A large workshop might employ designers, modellers, chasers and engravers. It also might employ women as burnishers, finishers and polishers. Some of these tasks could be subcontracted out to specialist workshops. The celebrity silversmith was an artist, but often also an entrepreneurial businessman.

During the nineteenth century, advances in technology, transportation and marketing led to the mass manufacturing of silverware for the upper and middle markets. Inventions such as the roller die, which could 'stamp' out shapes, facilitated large-scale production of items like cutlery. By the late nineteenth century, large manufacturing companies advertised a dizzying array of silverware through trade catalogues. Bridal couples in particular began to receive an abundance of silver at ever-more-commercialized weddings, the invention of plating methods making silver almost commonplace. Makers of luxury goods understood that their business was the manufacture of desire as well as of material objects, and that the injection of a little status anxiety further boosted sales. By the late nineteenth century, popular magazines could speculate that most households, except the most needy, boasted a few items of silver, or at least silver plate.

The prevailing zeitgeist was 'conspicuous consumption', a term coined by the American economist Thorstein Veblen in 1899 to explain ostentation as a hyper-competitive jockeying for place. The higher one's rung, the more one could afford to value the 'conspicuously wasteful', such as a solid silver spoon rather than one made of silver plate.[13]

His comparison was a timely one. For centuries artisans had been manipulating materials to imitate more costly ones. For example, Chinese metalworkers had been using a silver-coloured

Reed & Barton child's spoon, 1890s, silver.

alloy of copper, nickel and zinc called *bai tong*, anglicized to 'Paktong' for centuries. In many cultures, objects might be coated with silver foil or a thin sheet of silver. An industrialized version of silver plate was invented in Sheffield, England, in the mid-eighteenth century by a cutler attempting to repair a knife. Consisting of a 'sandwich' of copper between two sheets of silver, it was about one-third of the price of solid silver. As the silver plate was created before the item was formed, Sheffield plate could be fashioned with the same dies as silver objects, and was quickly considered a fashionable and innovative alternative, in London as well as northern England. A further revolution occurred with the invention of electroplated nickel silver (EPNS), patented in Britain in 1840. While Sheffield plate relied on heat and pressure, EPNS involved a chemical reaction using electricity to split a solution of silver salts, depositing a coating of silver on an alloy of copper, nickel and zinc. Much smaller amounts of silver, compared to the cheaper base metal, were required to form EPNS, making it an economic, easily produced and widely used alternative for silverware.

As silver became commonplace, it started to lose its sheen. Status symbols are notoriously fickle, and shift with each new wave of ownership. While Victorian newlyweds were bestowed with silver, the most popular wedding presents in today's food-centric culture are kitchenware and specialized culinary equipment, as well as cash, newly acceptable in the form of gift cards (in other cultures, of course, money has long been de rigueur as a wedding gift). Silver still occupies a niche, but an increasingly artisanal one.

Body conscious

As Emma Bovary discovered, the more something is unobtainable or forbidden, the greater its attraction. Sumptuary laws, which restricted consumption of luxury items (often dress), were created to regulate trade, maintain social hierarchies and protect scarce resources. They achieved these goals with various degrees of success, but what they were perhaps best at was stimulating appreciation for the forbidden item. From ancient Greece to Ming Dynasty China, to Elizabethan England, people hankered after gold jewellery, embroidered robes, big mausolea and the specific colour of purple they were told they couldn't have. In fourteenth-century Italy, what they craved were big silver belts.

Italian ambition

Italy became fashion-forward in the twentieth century with the founding of houses whose products we still covet – Gucci, Armani, Prada – but 'the first age of fashion' was in the fourteenth century. Fashion as a means of expressing status became a widespread focus of urban life, not simply of court life.[14] Ours is not the only era to pursue recreational shopping with a passion; in the major trading cities of fourteenth-century Italy, shoppers could browse through a marketplace of domestic and imported luxuries that would still be enticing today – exotic textiles from central Asia, Mediterranean wool finely finished in Florence, silk loomed in Lucca, and silver buttons, belts and hair ornaments crafted by master goldsmiths, whose main stock in trade was actually silver.

The desire to attract attention through dress, to impress neighbours and colleagues, to emulate – or better still, inspire trends – and to shop as a pastime, are all impulses we've shared over centuries. We all participate in the spectacle of fashion. What was different in fourteenth-century Italy was that it was not young people or women who were the trendsetters and arbiters of taste, but mature men – men not just of rank, but of wealth and civic status, and often the very men responsible for crafting the sumptuary laws that restricted the fashion privileges they enjoyed.

Dress became more revealing – in every way. Set-in sleeves and buttons meant that clothing could hug and shape the body in previously unsuspected ways. Fine fabrics, trims and accessories of precious metals and gems made possible a more immediate and blatant revelation of wealth than the plate on one's table. Inspired by military dress, men began to swagger in ensembles with daggers, flashy lengths of chain and the must-have accessory

– the silver belt. Made of broad links, and often set with gem-stones or enamelled plaques, it was slung over the hips in an ostentatious display of wearable wealth. No less a celebrity than Marco Polo willed several silver belts to his heirs in 1324.[15]

Such display was deemed appropriate for men of position, but for women, youth and children, it was an unseemly challenge to decorum and hierarchy, hence the spate of sumptuary laws that restricted both type and value of ornament, effectively limiting the amount of silver that could be worn by women and youth. This is not to say that they complied – records from several cities reveal numerous instances of women fined for wearing heavy silver belts and hair ornaments, or for carrying children with too many silver buttons on their sleeves.

The goldsmiths who fashioned these belts and ornaments either as bespoke or ready-made items became ever more skilled at creating the *impression* of weightiness with less silver. For instance, thin silver wire could be worked into an ornamental lattice, or silver foil could be applied to leather. A growing fashion for silver and gold thread woven into textiles or used for embroidery necessitated quantities of foil that was wrapped around silk thread. The laborious task of beating precious metals commonly fell to goldsmiths' apprentices, and there was close collaboration between the goldsmithing and textile industries in the production of luxury goods.

Over time, tempting new ways to display these goods evolved. Goldsmiths won the right to retail their goods at their own stalls in open-air markets. They appeared on the second-hand market, as fresh fashions forged novel designs. They could also be found on bankers' stalls, as they were understood to retain inherent worth as capital. And display – whether on the body or in the market – stimulated demand and the desire to imitate.

Silver threads

Aspiration is a universal impulse, as basic, perhaps, to our nature as breathing, to which it is related in its etymology. In fourteenth-century Italy, men of stature were found worthy of emulation, and silver belts and ornaments the objects of desire. In dynastic

China, it was the emperor and his court, and the dragon robe. Although dragon robes had appeared as early as the Tang Dynasty, they became more firmly established as court dress in the Yuan Dynasty (1279–1368) under the Mongols, and their use increasingly codified during the Ming Dynasty (1368–1644). Colours were restricted to certain groups (yellow for the emperor, red for high officials, blue for lower ranks and black for attendants), as were the number of claws on the dragons (five for the emperor and high-ranking princes, four for noblemen and senior officials).[16] The dragons were often embroidered in gold or silver thread, made by wrapping foil around a core of silk thread. These metal threads were then laid on the surface of the silk robes and 'couched' in place using silk thread.

Fashion, as the officials responsible for Italian sumptuary laws found out, can be a provocative and destabilizing force. Despite laws forbidding individuals from ordering their own, dragon robes became more widely fashionable during the early sixteenth century. Sewn from voluminous metres of silk, and glistening with silver dragons and the sheen of satin-stitch embroidery, these robes symbolized the ideal of the scholarly and courtly lifestyle to which their owners aspired. As the Qing Dynasty (1644–1911) wore on, restrictions on the number of dragon's claws were widely ignored, and most dragons sported five claws, regardless of the station of their wearers.

In China a lowly provincial official could emulate an emperor through the decoration of his robe; in Italy a young man might imitate a wealthy banker through a goldsmith's skilful application of silver foil. In India a woman might purchase a sari with synthetic metallic threads imitating the rich silver and gold zari work of traditional Banarasi saris. Authentic zari work decorated with threads of precious metals was historically affordable only to wealthy families. The silver- and gold-wrapped silk threads were woven into the fabric as part of the weft, creating a luscious, heavy textile that was sold by weight. In addition to woven metal threads, the textile might be decorated with applied silver foil and hammered spangles. A bride might receive a precious stash of such saris as part of her trousseau, collected over

time by her family, and supplemented by gifts from her new in-laws. Folded in a locked storage cabinet and kept for special occasions like weddings and religious festivals throughout her life, these saris embodied the status of both families. But they were also stores of wealth. Burning a sari to salvage the silver has been practised since Mughal times.

Most of the silver objects that ever existed are now lost to us. Large silver belts might have been all the rage in fourteenth-century Italy, but almost none survive. Sometimes they became coins, sometimes new objects. The survivors seem all the more precious, and their trajectories through history sometimes remarkable. The Victoria and Albert Museum owns a mag-nificent mid-eighteenth-century 'mantua'. Made of silk and embroidered with an extravagant 5 kg (11 lb) of silver thread, it would likely have been worn by a member of the aristocracy at the Georgian court, where a scarlet gown with 'great roses not unlike large silver soup plates' was noticed at a royal birthday.[17] More closely resembling a large piece of upholstered furniture rather than a dress, its extravagant width – 1.8 m (6 ft) across

Male court robe, China, 1850–70, silk twill with gold and silver metallic couching and silk embroidery.

at its widest – provides a spacious field for the riotously fecund 'Tree of Life' embroidery. Like Banarasi saris, these gowns were prized for their precious metal and not infrequently burnt to reclaim it. We can't know why this one survived; possibly its owner prized the skill of its embroidery. A signature on the train links it to a Huguenot house of embroiderers who had set up business in London, like other talented Protestant artisans escaping a repressive regime in France. Its subsequent history is a mystery until the 1920s, when it was bought to wear to a fancy dress ball.

An English mantua gown, 1740–45 (altered in the 1920s), silk, silver thread, silver spangles.

Because of its girth and weight, the mantua was a cumbersome piece of apparel, yet for those jostling for position at court, a necessary sign of status. Displaying status may well entail beauty and pleasure, but it can also literally be a burden. Silver is a heavy metal.

8 Pure

In Stephen King's novella *Cycle of the Werewolf*, the small Maine town of Tarker's Mills is ravaged each full moon by 'something inhuman'. As the months pass, the monster beheads a child, devours a lonely virgin, tears open a police constable's throat and exsanguinates a vagrant. In July the beast is temporarily halted by a disabled child. Trapped in his wheelchair, ten-year-old Marty resorts to flinging a firecracker into his attacker's face. On New Year's Eve it returns, but this time the child is ready with a pistol loaded with two pure silver bullets. When the werewolf lunges, Marty shoots twice, and defeats the demon with silver.[1]

The idea that silver is powerfully apotropaic has been embraced over centuries and across continents. Popular culture's recent fascination with all things vampiric deployed motifs of silver crosses, chains and bullets that barely required explanation, so deeply rooted is the notion that silver defeats evil.

It is a truism that warrants some decoding. Stephen King packs a folkloric punch in that brief image of Marty discharging his gun from his wheelchair. The hunters of demons have always been set apart in some way. And as a child, Marty is both especially vulnerable (werewolves lust after young blood), and powerful in his innocence. Vampires and werewolves – in some cultures these two creatures are one and the same – are mortals transformed into supernatural entities because they lived lives of violence, or were wronged. Those who defeat the bloodthirsty beasts restore proper order, release the victims and cleanse the community. Tarker's Mills' werewolf is actually the blameless

Baptist church minister whose misfortune was to pick some strange, short-lived flowers by a cemetery. 'I'm gonna try to set you free', Marty says as the werewolf leaps towards him.

Marty's bullets have been made by melting his silver confirmation spoon. In the Christian tradition, confirmation, like baptism, is a cleansing rite. The spoon given as a gift to mark the occasion is literally and symbolically 'pure'. Lead bullets might stop human vice, but demon slaying has nothing to do with velocity or the rupturing of vital organs. Silver vanquishes evil because of its purity.

Purity, however, is always Janus-faced. Pure cannot exist without impure, and the story of silver's purity is also that of its pollution. The undoing of Judas Iscariot is a poignant example. Vampires stalked the ancient world, and from the viewpoint of Christianity, Judas was among their brethren.[2] Corrupted by thirty pieces of silver, Judas betrays Christ with – of all vampiric gestures – a kiss. Repenting too late, Judas attempts to return the silver, flings the coins to the temple floor and hangs himself. As the coins are 'blood money', the priests are unwilling to return them to the treasury, and instead use them to buy a burial ground.

The ensuing legends depict Judas's inability to truly die, because he is doomed to exist forever without forgiveness. His silver-blighted story has echoed down the centuries through folklore, literature and the media, cropping up in the most unlikely places. In *The Judas Coin* (2012), by the veteran comic-book writer and artist Walter Simonson, one of the biblical silver coins reappears in episodes separated by centuries, wreaking havoc on the lives of those who discover it – all characters from the DC Comics universe. Embroiled in the chaos are Roman gladiators battling treachery in Germania, Caribbean pirates, card sharks in the American Wild West, Batman fighting museum thieves in Gotham City and deceitful anime girls who are eventually blasted into oblivion on an asteroid. While the silver coin is an irresistible lure, it also brings deception, disappointment and even death.

The symbolic purity of silver is closely bound to its material purity. Although appearances can be deceiving (over the ages,

Simon Bening, *Judas Receiving the Thirty Pieces of Silver*, Flemish, 1525–30, tempera colours, gold paint and gold leaf on parchment.

many white metal alloys have been passed off as silver), the way silver *looks* contributes a great deal to what it *means*. We would never, for instance, entertain any idea of the symbolic purity of tin or lead. These are, of course, more common metals and lack the cachet of rarity, but more importantly their appearance does not signal the qualities that we have come to associate with purity. The visual markers of silver's purity lie in its distinct physical properties. The surface of polished silver is lustrous white – a colour long associated in many cultures with pristine condition. Furthermore, as the most reflective of metals, polished silver seems to emanate light, and light has been a shared symbol for purity and goodness across time and cultures.[3]

Sacred silver

Many religions have drawn a connection between silver and sanctity. Among the earliest surviving silver objects from Anatolia are ones that appear to have had a religious function. The Hittites, who built a far-reaching empire based in Anatolia during the second millennium BCE, created silver vessels in the shapes of animals as religious offerings. After the Hittite king Hattusili I conquered the city of Zalpa on the Black Sea, his annals record that he gave a silver bull and a silver fist to the temple of the Storm God.[4] Ceremonial Hittite drinking vessels in the shapes of, for instance, fists and stout-necked bulls with flaring nostrils survive in museum collections. The Old Testament Book of Exodus places the receiving of the Ten Commandments and the pro-hibition against worshipping gods of silver and gold around the same time, suggesting a wider context in which this was prac-tised (Exodus 20:23). While the tribe of Israel was forbidden to worship silver idols, they were permitted to use precious metals in the adornment of religious edifices. According to the biblical description, Solomon's Temple at Dura Europos in present-day Syria was amply outfitted in both silver and gold, with silver tables, lampstands and dishes (1 Chronicles: 28:14–17).

The Romans made ample use of silver in religious rites. The Berthouville Treasure, discussed in the previous chapter, was

Vessel terminating
in the forepart of a
bull, Hittite, 13th–14th
century BCE, silver.

dedicated to the Gallo-Roman god Mercury. Although most of the vessels were originally for domestic use and only later offered as votive gifts at the sanctuary, two statues of the god were clearly religious. The most sumptuous is a muscled nude with tightly curled hair. At half a metre (20 in.) tall, it is the largest repoussé silver statue to have survived from the Roman era, expertly crafted from six hammered and joined plates of silver.[5] As the god of commerce, Mercury was frequently formed from silver and depicted holding a bag of money. The donor who commissioned this statue may have wished to acknowledge Mercury's benevolence in his or her own life – the impressive scale of the statue and the purity of the silver suggest a very wealthy patron.

Roman silversmiths depicted not only gods but their temples. The Apostle Paul ran afoul of this artistic expression in Ephesus (in present-day Turkey) during a mission trip in the middle of the first century. His evangelism, with its anti-idolatry rhetoric, threatened the handsome livelihoods of craftsmen who sold small silver statues of the goddess Artemis' temple – a wonder

Statuette of the
Roman god Venus,
2nd–3rd century, silver.

of the ancient world. The affront to Roman piety (and tourism) resulted in a riot (Acts 19:23–9).

The early Christian Church was hardly characterized by wealth and expenditure on precious metals. However, after the official recognition of Christianity by the Roman emperor Constantine in the early fourth century, the Church began to enjoy generous patronage. As it matured in a Roman world richly endowed with silver, Christianity transformed the Roman meanings of silver into its own sacred ones.

It is clear from biblical references that silver already had a host of metaphorical meanings in Judaic culture. Although most references to silver pertained to money and wealth, there was an early equation between the clean, white metal and divine purity. According to the Psalmist, 'the words of the Lord are flawless, like silver purified in a crucible' (Psalm 12:6). Ancient metallurgy proved a treasure trove for Israel's poets. Through travail, God's people, too, were purified, as the Psalmist records, 'you, God, tested us; you refined us like silver' (Psalm 66:10). The Christian Church inherited these figures of speech, and made them visible through the act of worship.

Most elite Roman families owned silver, and took pleasure in its display and use.[6] Gifts of silver were part of the fabric of Imperial Rome; at the highest level, emperors bestowed sumptuous inscribed *largitio* (ceremonial gift) dishes on favoured officials and soldiers. Following Christianity's imperial recognition, the Church became the recipient of lavish endowments, often in precious metals. Constantine gave 5,000 kg (11,025 lb) of silver to the Lateran Basilica, the first Christian basilica in Rome.[7] Churches became treasure houses, glittering with silver-clad altars, doors, chancel screens and liturgical vessels.

Given the lavish use of silver at the Roman table, it was apposite that the bread and wine of the Christian Eucharist should be served from silver. The most sacred vessels of early Christianity were the paten (the platter on which the Eucharistic bread was served) and the chalice for the wine. At first, these sacred liturgical vessels were more highly prized than any images.[8] The typical form for the paten was a large, deep platter with a flat rim. It was simply decorated, often with the *Chi Rho* monogram (made by superimposing the first two capital letters of the Greek word for Christ: ΧΡΙΣΤΟΣ). And, as with Roman votive gifts, it was often inscribed with the donor's name. A hoard of fourth-century liturgical silver, which may have been used for the celebration of the Eucharist, was found after ploughing a field at Water Newton, Cambridgeshire, in 1975, and is now held in the British Museum. Many of the pieces have the *Chi Rho* symbol, and many are dedicated, such as a bowl

Dove from the Attarouthi Treasure, Byzantine (Syria), 500–650 CE, silver. This collection of silver liturgical objects once belonged to a church in Attarouthi.

whose inscription reads, 'I Publianus, honour your sacred shrine, trusting in you, O Lord.'[9] Other bowls were dedicated by the female donors Amcilla, Viventia and Innocentia.

The paten and chalice were the most important liturgical vessels but, similar to silver dining sets, soon a plethora of specialized silver objects was pressed into service: ladles, pierced spoons to strain wine, candlesticks, fans to swipe any importunate flies off the host and containers for the bread. Silver proliferated in sanctuaries, covering apses, tombs and gospel books. By the fourth century, the cult of Christian martyrs was in full bloom, and churches eagerly sought out the remains of saints associated with their cities. Sumptuous containers were crafted in the forms of the body parts they preserved. For centuries afterwards, reliquaries in silver and gold, ornamented with precious stones, housed collections of bones, skulls and teeth.

While the interiors of churches gleamed and the dead were housed in silver, outside of those walls the homeless and impoverished gleaned whatever scraps of charity might fall their way. Cities such as Antioch and Constantinople had splendidly endowed churches and worldly, free-spending Christians, but also needy widows, orphans and indigents. The most vociferous criticism of the profligacy of fourth-century Christians came from the upper echelon of the Church itself. John Chrysostom, first as a priest in Antioch, and then, from 398 CE, as the archbishop of Constantinople, railed against the hedonism of those who splurged on silver dining sets, jewellery and ivory beds decorated in silver, while their brothers starved. Chrysostom had been conditioned to austerity by spending several years as an ascetic monk. Back in a metropolis burgeoning with temptation, he seems to have been particularly affronted by the citizens' lavish spending on silver. 'Golden Mouth', as the Church leader was known due to his eloquence, could also be cuttingly earthy:

Arm reliquary, France, 13th century with 15th-century additions, silver, silver-gilt, glass and rock crystal over wood core.

'one man defecates in a silver pot, another has not so much as a crust of bread.'[10]

The silver chamber pot was a malodorous target for critics of Christian vice. In an episode from the seventh-century Greek text *The Life of St Theodore of Sykeon*, the saint sends his archdeacon to Constantinople to purchase a silver paten and chalice. On the first occasion the mass is celebrated with the new silver, the ware mysteriously turns black with tarnish. An investigation reveals that the sacred vessels had been made from the melted silver chamber pot of a prostitute.[11] While silver is pure, it can also be polluted; it can symbolize both the sacred and the sacrilegious. The epistolist James had a warning for those with wealth but no charity: 'Your gold and silver are corroded. Their corrosion will testify against you and eat your flesh like fire' (James 5:3). Chrysostom, in one of his most poignant comments on luxuries such as silver, observes, 'They were leaves – winter seized them, and they all withered up. They were a dream – and when day came, the dream departed.'[12]

The vanity of material possessions, and by extension the ephemerality of the material world, is a key concept too of Buddhism. In the fourth-century text the *Diamond Sutra*, the Buddha tells his disciple that even if a person were to give away silver, gold and gems, this would be inconsequential beside memorizing and sharing one stanza of Buddhist teaching. Chrysostom and Buddha were worlds apart, yet they shared the metaphor of the dream. In the *Diamond Sutra*, after emphasizing the inefficacy of giving silver to deities, the Buddha calls the entire fleeting world 'A star at dawn, a bubble in the stream, a flash of lightning in a summer cloud, a flickering lamp, a phantom, and a dream.'[13]

The Buddha achieved enlightenment about the impermanence of things while sitting under a bodhi tree. Ironically, the Mahabodhi Temple later constructed on the site was a celebration of the sumptuous. The temple complex, situated in the modern state of Bihar, was described in the writings of the Chinese monk Xuanzang, who visited it during a sixteen-year pilgrimage to India in the seventh century. He records doors and windows encased in silver and gold, and studded with pearls and gems, as well as solid silver statues 3 metres (10 ft) tall and located in large niches.[14]

The Buddha's words urged followers to eschew worldly luxuries, but throughout lands where Buddhism spread, artisans crafted devotional objects in precious metals for temple and personal use, and the Buddha's words, urging detachment from earthly treasure, might themselves be written down in silver. Buddhism spread to China and developed there in a unique form. In hundreds of Mongolian monasteries, thousands of scribes painstakingly copied Buddhist texts. Production burgeoned from the sixteenth century, when the Mongols began producing their own paper. In the eighteenth and nineteenth centuries, the most lavish were produced on black paper with silver or gold ink.[15]

Seated Buddha, Korea, 13th–14th century, cast silver with gilding.

The art of manuscript illumination flourished during different time periods in Buddhist, Christian and Islamic art. The purity of the sacred words was emphasized through the pure materials of silver and gold. In the Islamic world, from Iraq to India and North Africa, volumes from the Koran were reproduced in elegant calligraphy accented with gold or silver. In rare, very luxurious cases, embellished silver or gold script sparkles like a nightscape on black or dark-blue paper. At the same time, Islam was as ambivalent to precious metals as Christianity and Buddhism. While the faithful are urged not to hoard silver and gold, and to practise charity, they are also promised a paradise where they wear silver bracelets and drink from silver goblets (Koran 9:34; 76:15; 76:21).

Across world religions, copying the sacred text was itself an act of worship which could lead to purification. In the Jewish

Friedrich August Ferdinand Eisolt, Torah crown, Germany, 1854–60, silver.

Anthony Elson, censer commissioned for Lincoln Cathedral, 2007, silver.

synagogue, reverence for the word was made explicit in the housing of the Torah scrolls. In the earliest days, these were carefully wrapped in cloth, but gradually the creation of a Torah case, or *tik*, provided silversmiths with the same opportunity for devotion that was afforded scribes. From the eleventh century onwards, *tikim* were made from elaborately worked silver, with ornamented staves and crowns. As it was forbidden to touch the scroll, silversmiths devised pointers, often in the shape of a pointing hand. The creation of silver housing was in obedience to the Talmudic command to 'beautify the law'; across world faiths, the shared acknowledgement that 'God is beautiful and loves beauty' led to decoration of the divine word and places of worship with the beautiful metal.[16]

Alchemical silver

Alchemy is a millennia-old experiment in purification. At its most materialistic, alchemy was the attempt to transform base metals into precious ones. An eclectic and often-arcane art, it was practised in ancient Greece, India, China, the Islamic world and Renaissance Europe, and is still a subject that fills booksellers' shelves, often (incongruously) under the rubric of 'New Age', and one that proliferates online. It was peddled by the charlatans of the medieval marketplace, exercised by protochemists and healers and applied by craftsmen in pursuit of beauty. During the Renaissance, alchemical knowledge had become a commodity that found patrons at the highest levels of society. By the mid-sixteenth century, the courts of the Holy Roman Empire were supporting alchemical laboratories generously stocked with ovens, crucibles, vials, instruments and chemical elements. These well-staffed laboratories were engaged in creating tinctures – the essential purifying substances – that would produce wealth from the worthless. Although alchemy has long been ridiculed as chicanery, in the early modern period it was considered of great practical and commercial value. As the productivity of central Europe's silver mines began to diminish in the late Renaissance, German and Bohemian princes invested in projects to boost their territories' mineral wealth, and alchemy seemed to promise a means of maximizing mining profits. Applying alchemical knowledge to the smelting process appeared to be a practical solution to the difficulty of extracting precious metals from unpromising ore. Metallurgy, alchemy and mining expertise intertwined, with princes, investors and mine owners establishing metallurgical or alchemical laboratories on mine sites. Investment in mining technology, improvements in the smelting process and greater knowledge of the composition of ores indeed boosted mine profits for a while, and patrons such as Duke Julius of Braunschweig-Wolfenbüttel, whose territory included mines in the Harz Mountains, were keen to credit alchemy for their success.[17]

Attractive as increased mine production might be, alchemy was also valued as a font of knowledge about the natural world. Far from being an obscure branch of investigation, alchemy is threaded throughout the Renaissance pursuit of learning. This was never more explicit than at the early seventeenth-century court of Emperor Rudolph II in Prague, which gathered together poets, astronomers, theologians, painters, inventors and two hundred alchemists who laboured in the palace's alchemical 'kitchens'.[18] It was not physical silver and gold that Rudolph II pursued, but knowledge. Rudolph's alchemical kitchens were one part of a cosmos that included a menagerie, botanical gardens, art and mineral collections, and mechanical objects. Rudolph II was ravenous for knowledge as a way to decode the secrets of nature. Just as practical alchemy sought to manipulate nature (for example, ore) to produce precious metal, organic alchemy pursued knowledge of healing and the purification of the body. The desired results were for improved health, vitality and youthful vigour – and even, in some strands of alchemy, such as of ancient China, immortality. At the very highest plane, alchemy endeavoured for an enlightened mind, a higher level of consciousness and a purified soul. From the beginning, alchemy's earliest practitioners understood that humanity itself was the subject of transformation. As Mircea Eliade, the Rumanian philosopher and chronicler of man's spiritual adventures explained, 'the Western alchemist, in his laboratory, like his Indian or Chinese colleague, worked upon himself – upon his psycho-physiological life as well as on his moral and spiritual experience.'[19] The base matter was individual consciousness, or the flawed self. Through the alchemical process, the soul was removed from the body and cleansed, which was likened to turning lead into silver. Carl Jung, who in the mid-twentieth century did much to elucidate the often fearsomely abstruse philosophy of alchemy, described this intermediate stage as 'the silver, or moon condition'.[20] The purified soul may then be reunited with the body to achieve the gold, or sun condition. The perfection of man, then, is explained through the transmutation of metals, from the base, through pure silver, to transcendent gold.

The notion that elements on earth were governed by the planets existed as early as the sixth century BCE. Silver and gold were understood to 'grow' under the influence of the moon and the sun. It was an obvious visual analogy, equating the bright disc of the moon to the clean sheen of silver, and the fiery sun to gold. The alchemical symbols of the moon and the sun are mercifully lucid signs in a frequently confounding art. To return, briefly, to the defeat of vampires and werewolves, silver is powerful not only because of its purity, but its association with the moon – the light in the darkness that can extinguish evil and lunacy.

Protection

Although silver bullets vanquish monsters, silver has long been worn on the body to deter illness, accident, danger or malice. Silver jewellery beautifies the wearer, is a store and advertisement of wealth, and sometimes has a practical function as a fastener. But across cultures, silver ornament has been understood – sometimes in very generalized terms – to bring good fortune. The earliest jewellery might well have had a protective as well as an aesthetic purpose; in addition to attracting admiring glances, personal adornments could also thwart the evil eye. Silver's shininess made it an especially effective choice as an amulet for deflecting harm.

Faith in amulets is as old as faith itself. In ancient Egypt, amulets decorated with auspicious symbols, or in the forms of animals or deities, were worn as bracelets, necklaces and rings,

Counterweight, Dong minority, Guizhou, China, silver.

Dong minority woman wearing a counterweight, which balances an apron worn at the front. Photograph taken in Zhaoxing village, Guizhou, China, 1997.

or placed in the wrappings of mummies. Across societies, it was recognized that times of transition – birth, adolescence, marriage, pregnancy and death – were times of great physical or spiritual vulnerability, which could be tempered by the magic of the amulet. Although belief in their power was often denounced as foolish or dangerous superstition, gifting an amulet was, above all, motivated by love and concern for safe passage.

Necklace with 'Hand of Fatima' motif, Palestine, 1880–1930, silver.

As expressions of faith, amulets have had an ambivalent relationship to orthodox religion. Sometimes they were condemned by religious authorities as idolatrous, but often they were tolerated, or an attempt was made to transform their 'pagan' orientation into an acceptable expression of piety. In 1979 two amulets in the form of tiny silver scrolls were discovered in a cave near Jerusalem. They dated from around the late seventh century BCE and had faintly scratched letters in Hebrew script. One read, 'Blessed be [he or she] by Yahweh, the helper and dispeller of evil. May Yahweh bless you [and] protect you, and may he cause his face to shine upon you and grant you peace.'[21] Judaic amulets were typically inscribed with sacred verses, and this practice continued into the Christian era. Amulets with God's name and Bible verses were highly fashionable in early Christian Palestine, Syria and Asia Minor, despite the Church's best efforts to suppress their use. The Council of Laodicea (from around 363) expressly forbade Church leaders to be sorcerers, conjurors or makers of amulets.

Amulets, similar to silver tableware, were quite specialized in their application. Different forms countered different maladies. Protection against the evil eye was perennially in vogue, and appears in different cultural forms as the Judaic 'Hand of Miriam' or the Islamic 'Hand of Fatima'. Amulets in the shapes of body parts could invoke healing for specific limbs, but perhaps the strangest was a 'tongue of St Nepomuk', a wax tongue cradled by silver filigree. St Nepomuk was a fourteenth-century Bohemian priest famous for his absolute discretion, and naturally his 'tongue' protected the wearer against lies and gossip.

Sometimes harm could come from another human being or natural causes; sometimes it had a supernatural origin. One of the

reasons amulets and charms were so censured by orthodox religion was because of their association with atavistic superstition. Norway became Christianized around 1000 CE, and specifically Lutheran from the sixteenth century. However, strong remnants of pre-Christian belief survived for centuries afterwards, particularly in rural areas. Norwegian oral legends were populated with a pantheon of ghosts, goblins, trolls, giants, witches and *huldrefolk* – underground creatures who assumed human form, and loitered with the intent of stealing babies, distracted children and vulnerable adults. Silver, which the Norwegians wore in the form of jewellery, collar pins and metallic lace, was considered a powerful deterrent.[22] One of the most popular forms of talisman was the *sølje*, a large silver brooch that measured up to 9 cm (3.5 in.) wide, often made of filigree floral motifs. Dangling ornaments which caught the light made the brooches even more effective at deflecting evil. Each time the brooch was worn in church, or passed down to the next generation, it assumed even greater protective power.

The opportunistic *huldrefolk* were more likely to abduct people who were alone, such as hunters or women working in remote pastures. Common to much folk belief, times of passage were particularly dangerous, and on festivals such as at midsummer and at weddings, women might pin three brooches to their costumes. The vulnerability of children – especially unbaptized babies – caused enough concern for them to have brooches pinned to their wrappings, or shiny knives and scissors placed in their cribs.[23]

Amulet of the tongue of St John Nepomuk, Germany, 1830–50, silver, wax and glass.

Woman wearing a silver brooch, Telemark, Norway, c. 1895.

Clean

Norwegian babies who survived these sometimes contradictory efforts to keep them safe were baptized. Fonts in rural churches were typically made of stone, but European aristocratic babies were more likely to be baptized in silver. The Swedish royal family, for instance, commissioned a silver baptismal font in 1696 for the new Royal Chapel, inside the Stockholm Palace, and it has been used to christen royal babies ever since. In Britain, the silver-gilt Lily Font was made for the baptism of the first daughter of Queen Victoria and Prince Albert in 1841. The event

Hallvard Bjorgulfson Straume, *Skålsølje* (bowl brooch), Norway, *c.* 1880, silver with gold plating.

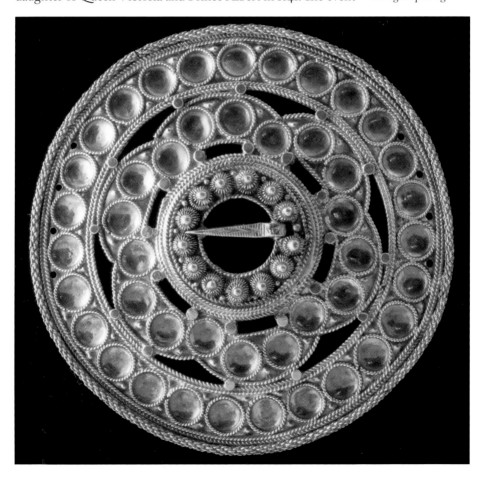

was laden with symbolism, from the 'purity' of the font's lily motifs, to the water from the River Jordan with which the Archbishop of Canterbury baptized the infant princess. Over the ensuing years, most British royal babies have been christened in the Lily Font.

Baptism is a cleansing rite, symbolizing the purification and renewal of the soul. Quite apart from its prestige as a precious metal, silver was suitable on a symbolic level for its purity. The associations between physical cleanliness and spiritual purity are long-standing. Cleanliness is next to godliness in many religious traditions, with ablutions playing a central part in preparation for coming before God in prayer. The medieval Islamic text *The Mysteries of Purity* explains the four stages of purification: bodily cleansing, purifying the senses of sinful urges, purification of the heart and finally purification of the self of everything but God.[24] External cleanliness is a prelude and foundation.

Major ablutions might require a visit to a bathhouse (*hammam*) to attain 'healthy cleanliness'. In the Middle East, a vibrant bathhouse culture traced its roots to Roman *thermae*. Although public baths later died out in Western Europe, in the east by the Middle Ages they had become an important social institution, integrated into complexes of mosques, schools and gardens. These *hammams* were raised to a level of steamy, fragranced perfection in the Ottoman world, serving to promote both piety and public health, and providing gathering places for both genders, though men and women were segregated. Bathing accessories became lavish, particularly at the upper strata of society. Attendants poured water from chased silver bowls, towels were intricately embroidered with coloured and silver threads, and female bathers made their careful way across the slick floors in high wooden clogs clad in sheets of repoussé silver.

Women might carry silver boxes in which to store jewellery and make-up. *The Mysteries of Purity* prescribes the application of eye kohl for men as part of cleansing rituals. Across cultures, the notion of cleanliness has encompassed grooming and make-up.[25] Probably it was early observed that perfumes, eye paints and lip balms kept better when stored in silver (today, one of the applications for nanosilver is the preservation of cosmetics). Silver flasks for cosmetics, dating from as early as the third millennium BCE, have been found in graves in Iran. In the Roman world, mosaics and wall paintings depicting the routines of the bathhouse sometimes portrayed attendants carrying silver boxes presumably stocked with small necessaries. Some of these boxes resemble the Muse Casket, a late Roman silver container for toiletry articles. The casket is in the form of a repoussé 'clamshell' fitted with five silver containers for perfumes and balms. It was probably made as part of a wedding gift for the Roman couple Projecta and Secundus, whose names are inscribed on another casket from the same treasure.

Decorating the sides are repoussé panels depicting the muses. Perhaps these were purely ornamental; perhaps they were meant in praise of the bride's own accomplishments. But these sinuous figures who shimmer against the pale surface embody for us the story of silver itself. There is Urania, muse of astronomy, a reminder of the exploding stars which birthed silver.

Cosmetic
or medicine box,
China, *c.* 700–800 CE,
silver, gilt decoration.

The Muse Casket,
330–370 CE, silver.

There is Clio, muse of history, recalling silver's world-shaping march across continents. There is Melpomene, muse of tragedy, reminding us of silver's cost in human suffering; Polyhymnia, muse of the sacred, reflecting our desire for the pure; and the muses of poetry, music and dance – reminders that silver is beautiful, and has given us pleasure for millennia.

REFERENCES

1 The Nature of Silver

1 Olympic Studies Centre, *Olympic Summer Games Medals from Athens, 1896 to London 2012* (2013), available at https://stillmed.olympic.org, accessed 25 July 2016.

2 Quoted by Margaret Mitchell, 'Silver Chamber Pots and Other Goods which Are Not Good: John Chrysostom's Discourse against Wealth and Possessions', in *Having: Property and Possession in Religious and Social Life*, ed. W. Schweiker and C. Mathewes (Cambridge, 2004), p. 107.

3 R. Wakshlak, R. Pedahzur and D. Avnir, 'Antibacterial Activity of Silver-killed Bacteria: The "Zombies' Effect"', *Nature Scientific Reports*, v/9555 (23 April 2015).

4 Robert Cook, 'Connoisseur's Choice', *Rocks and Minerals*, LXXVIII (January/February 2003), p. 45.

5 Ibid., pp. 43–4.

6 C. J. Hansen et al., 'Silver and Palladium Help Unveil the Nature of a Second R-process', *Astronomy and Astrophysics*, DXLV (2012).

7 Thomas Goonan, *The Lifecycle of Silver in the United States in 2009*, U.S. Geological Survey Scientific Investigations Report (2013), p. 1, available at https://pubs.usg.gov, accessed 25 July 2016.

8 I am grateful to Dana Willis, Director, Resource Geology, Coeur Mining, for discussions on exploration and sources on which I drew for this section.

9 *Anatomy of a Mine from Prospect to Production*, U.S. Department of Agriculture Forest Service Technical Report (February 1995), p. 30, available at www.fs.fed.us, accessed 25 July 2016.

10 I would like to thank Dana Willis for this resource.

11 C. Graham and V. Evans, 'The Evolution of Shaft Sinking Systems in the Western World and the Improvement in Sinking Rates', *Canadian Institute of Mining, Metallurgy and Petroleum Magazine* (August 2007).

12 'Lucky Friday #4 Shaft Project', www.hecla-mining.com, accessed 23 September 2015.
13 Philippa Merriman, *Silver* (London, 2009), p. 72.
14 The Assay Offices of Great Britain, *Hallmarks on Gold, Silver or Platinum* (London, n.d.).
15 Susan Mosher Stuard, *Gilding the Market: Luxury and Fashion in Fourteenth-century Italy* (Philadelphia, PA, 2006), pp. 151–3.

2 Silver Landscapes

1 Martin Rees, 'We're the "Waste" from Distant Stars', *The Guardian* (1 May 2008), www.theguardian.com, accessed 25 June 2016.
2 N. H. Gale and Z. A. Stos-Gale, 'Cycladic Lead and Silver Metallurgy', *Annual of the British School at Athens*, LXXVI (1981), p. 176.
3 Herodotus, *The History of Herodotus*, Book III, section 57, trans. G. C. Macaulay, available at www.gutenberg.org, accessed 29 July 2015.
4 G. Jones, 'The Roman Mines at Riotinto', *Journal of Roman Studies*, LXX (1980), p. 146; Barry Yeoman, 'The Mines that Built Empires', *Archaeology*, LXIII/5 (September/October 2010), p. 23.
5 Jiri Majer, 'Ore Mining and the Town of Joachimsthal/Jachymov at the Time of Georgius Agricola', *GeoJournal*, XXXII/2 (February 1994), p. 91.
6 Quoted in J. Reid and R. James, eds, *Uncovering Nevada's Past: A Primary Source History of the Silver State* (Reno, NV, 2004), p. 48.
7 Mark Twain, *Roughing It* (San Francisco, CA, 1872), p. 378.
8 Geoffrey Blainey, *The Rush that Never Ended: A History of Australian Mining* (Melbourne, 1963), p. 155.
9 Mircea Eliade, *The Forge and the Crucible: The Origins and Structures of Alchemy* (New York, 1962), p. 57.
10 Nomination dossier for the Iwami Ginzan Silver Mine and its cultural landscape for inscription on the World Heritage List, presented to UNESCO, 2007, p. 31, available at http://whc.unesco.org.
11 Tony Waltham, 'The Rich Hill of Potosi', *Geology Today*, XXI/5 (September–October 2005), p. 187.
12 This and subsequent quotes are from *The Devil's Miner*, documentary film directed by Kief Davidson and Richard Ladkani, 2005.
13 Heraclio Bonilla, 'Religious Practices in the Andes and Their Relevance to Political Struggle and Development: The Case of El Tío and Miners in Bolivia', *Mountain Research and Development*, XXVI/4 (November 2006), p. 336.
14 Victor Montoya, 'The Tío of the Mine', trans. Elizabeth Miller (2008), www.margencero.com, accessed 22 June 2015.

15 Susan Deering, *Images of America: Silverado Canyon* (Chicago, IL, 2008), p. 47.
16 My thanks to Alan Gallegos, Southern Sierra Province Geologist, Sequoia National Park, for discussions on the occurrence of silver in southern California.
17 Richard Lingenfelter, *Bonanzas and Borrascas: Gold Lust and Silver Sharks, 1848–1884* (Norman, OK, 2012); this book offers an extensive overview of mining stock speculation at this time.
18 Ibid., p. 251.
19 Robert Louis Stevenson, *The Silverado Squatters* (London, 1883), e-book location 1518.
20 Deering, *Images of America*, p. 55, referencing stock certificate book at the Bowers Museum.
21 Yeoman, 'Mines that Built Empires', p. 24.
22 'Carson River Mercury Site', www.yosemite.epa.gov, accessed 1 July 2015.
23 Coeur Mining, *Corporate Responsibility Report* (2013), pp. 20–21.
24 'Mine Closure Done Right', www.barrick.com, accessed 1 July 2015.

3 Silver Transformed

1 Beryl Barr-Sharrar, 'A Silver Triton Handle in the Getty Museum', *Studia Varia from the J. Paul Getty Museum*, 1 (*Occasional Papers on Antiquities*, VIII) (1993), p. 99.
2 Clare Le Corbeiller, 'Robert-Joseph Auguste, Silversmith – and Sculptor?', *Metropolitan Museum Journal*, XXXI (1996), pp. 211–18.
3 I would like to thank Tal Arshi of Yemenite Art for this information.
4 Ester Muchawsky-Schnapper, 'An Exceptional Type of Yemeni Necklace from the Beginning of the Twentieth Century as an Example of Introducing Artistic Novelty into a Traditional Craft', *Proceedings of the Seminar for Arabian Studies*, XXXIV (2004), p. 181.
5 'Silver Disc Brooch of Aedwen', www.britishmuseum.org, accessed 10 July 2015.
6 Dexter Cirillo, *Southwestern Indian Jewelry* (New York, 1992), p. 97.

4 Empire Building: Two Coins that Changed the World

1 Nicholas Doumanis, *History of Greece* (London, 2010), p. 24.
2 C.J.K. Cunningham, 'The Silver of Laurion', *Greece and Rome*, XIV/2 (October 1967), pp. 145–6.
3 Errietta Bissa, 'Investment Patterns in the Laurion Mining Industry in the Fourth Century BCE', *Historia: Zeitschrift für Alte Geschichte*, LVII/3 (2008), pp. 263–73.
4 Peter Levi, *Atlas of the Greek World* (Oxford, 1984), p. 113.

5 T. E. Rihll, 'Making Money in Classical Athens', in *Economies Beyond Agriculture in the Classical World*, ed. D. Mattingly and J. Salmon (London, 2002), p. 134.

6 Ibid., p. 116.

7 Ibid., p. 118.

8 John H. Kroll, 'What About Coinage?', in *Interpreting the Athenian Empire*, ed. John Ma, Nikolaos Papazarkadas and Robert Parker (London, 2009), p. 89.

9 John Ma, 'Afterword: Whither the Athenian Empire?', in *Interpreting the Athenian Empire*, p. 223.

10 Simon Hornblower, *The Greek World, 479–323 BC* (London, 1983), p. 12.

11 Quoted in Doumanis, *History of Greece*, p. 28.

12 Aeschylus, *Persians*, trans. Herbert Weir Smyth, line 237, www.perseus.tufts.edu, accessed 3 August 2015.

13 Owen Jarus, 'Athenian Wealth: Millions of Silver Coins Stored in Parthenon Attic', www.livescience.com, accessed 3 August 2015.

14 Thucydides, *The Peloponnesian War*, 11/41, www.perseus.tufts.edu, accessed 5 August 2015.

15 Larry Allen, *The Encyclopedia of Money* (Santa Barbara, CA, 2009), p. 107.

16 Neil MacGregor, *A History of the World in 100 Objects* (London, 2010), p. 517.

17 Henry Kamen, *Golden Age Spain* (London, 2005), pp. 25–6.

18 Kendall Brown, *A History of Mining in Latin America from the Colonial Era to the Present* (Albuquerque, NM, 2012), p. 5.

19 William S. Maltby, *The Rise and Fall of the Spanish Empire* (New York, 2009), pp. 53–4.

20 Ibid., p. 57.

21 Brown, *History of Mining*, p. 7.

22 Quoted ibid., p. 16.

23 Niall Ferguson, *The Ascent of Money: A Financial History of the World* (New York, 2008), p. 23.

24 Maltby, *Rise and Fall*, p. 67.

25 Gwendolyn Cobb, 'Supply and Transportation for the Potosí Mines, 1545–1640', *Hispanic American Historical Review*, XXIX/1 (February 1949), p. 30; Thomas Cummins, 'Silver Threads and Golden Needles: The Inca, the Spanish, and the Sacred World of Humanity', in *The Colonial Andes: Tapestries and Silverwork, 1530–1830*, ed. Elena Phipps, Johanna Hecht and Cristina Esteras Martín (New York, 2004), p. 3.

26 Quoted in Cristina Esteras Martín, 'Acculturation and Innovation in Peruvian Viceregal Silverwork', in *The Colonial Andes*, p. 61.

27 Cummins, 'Silver Threads', p. 11.

28 Kris Lane, 'Potosí Mines', *Latin American History: Oxford Research Encyclopedias* (May 2015), http://latinamericanhistory.oxfordre.com.

5 Rivers of Silver from the New World to the Middle Kingdom

1 Quoted in Cristina Esteras Martín, 'Acculturation and Innovation in Peruvian Viceregal Silverwork', in *The Colonial Andes: Tapestries and Silverwork, 1530–1830*, ed. Elena Phipps, Johanna Hecht and Cristina Esteras Martín (New York, 2004), p. 61.

2 Harry Kelsey, *Sir Francis Drake: The Queen's Pirate* (New Haven, CT, 1998), pp. 148–57.

3 Ibid., pp. 46–50.

4 William S. Maltby, *The Rise and Fall of the Spanish Empire* (New York, 2009), p. 84.

5 Henry Kamen, *Golden Age Spain* (London, 2005), p. 42.

6 Niall Ferguson, *The Ascent of Money: A Financial History of the World* (New York, 2008), p. 26.

7 Quoted in Kamen, *Golden Age Spain*, p. 42.

8 Charles C. Mann, *1493: Uncovering the New World Columbus Created* (New York, 2011), p. 154.

9 Dennis O. Flynn and Arturo Giráldez, 'Born with a "Silver Spoon": The Origin of World Trade in 1571', *Journal of World History*, VI/2 (Autumn 1995), pp. 201–21.

10 Richard von Glahn, *Fountain of Fortune: Money and Monetary Policy in China, 1000–1700* (Berkeley, CA, 1996), p. 25.

11 Flynn and Giráldez, 'Born with a "Silver Spoon"', p. 208.

12 Dennis O. Flynn and Arturo Giráldez, 'Cycles of Silver: Global Economic Unity through the Mid-eighteenth Century', *Journal of World History*, XIII/2 (autumn 2002), p. 393.

13 Arturo Giráldez, *The Age of Trade: The Manila Galleons and the Dawn of the Global Economy* (Lanham, MD, 2015), p. 1.

14 Flynn and Giráldez, 'Born with a "Silver Spoon"', p. 201.

15 Ibid., p. 205.

16 Lucille Cha, 'The Butcher, the Baker, and the Carpenter: Chinese Sojourners in the Spanish Philippines and their Impact on Southern Fujian (Sixteenth–Eighteenth Centuries)', *Journal of the Economic and Social History of the Orient*, XLIX/4 (2006), p. 515.

17 Ibid., p. 520.

18 Mann, *1493*, p. 150.

19 Ibid., p. 153.

20 Flynn and Giráldez, 'Cycles of Silver', p. 398.

21 Katharine Bjork, 'The Link that Kept the Philippines Spanish: Mexican Merchant Interests and the Manila Trade, 1571–1815', *Journal of World History*, IX/1 (spring 1998), p. 42.

22 James Creassy, letter to John Baker Holroyd, 8 November 1804, quoted in Richard H. Dillon, 'The Last Plan to Seize the Manila Galleon', *Pacific Historical Review*, xx/2 (May 1951), p. 124.

23 Giráldez, *The Age of Trade*, p. 152.

24 Hugh Thomas, *World without End: The Global Empire of Philip II* (London, 2014), p. 255.

25 Quoted in William Lytle Schurz, 'Acapulco and the Manila Galleon', *Southwestern Historical Quarterly*, xxii/1 (July 1918), p. 31.

26 Ibid., p. 32.

27 Flynn and Giráldez, 'Born with a "Silver Spoon"', p. 214.

6 The New Flow of Demand

1 Sam Kean, *The Disappearing Spoon: And Other True Tales of Madness, Love, and the History of the World from the Periodic Table of the Elements* (2010), kindle ebook location 2099.

2 The Silver Institute, *World Silver Survey 2015: A Summary* (London, 2016), p. 11.

3 Jane Hayward, *English and French Medieval Stained Glass in the Collection of the Metropolitan Museum of Art* (New York, 2003), p. 249.

4 Jane E. Boyd, 'Silver and Sunlight: The Science of Early Photography', *Chemical Heritage* (Summer 2010), www.chemicalheritage.org, accessed 7 September 2015.

5 United States Geological Survey (USGS), *The Lifecycle of Silver in the United States in 2009* (Reston, VA, 2014), p. 12.

6 J. Wesley Alexander, 'History of the Medicinal Use of Silver', *Surgical Infections*, x/3 (2009), p. 289.

7 Arthur Williams, 'Alfred Barnes, Argyrol and Art', *Pharmaceutical Journal* (23 December 2000), www.pharmaceutical-journal.com, accessed 9 December 2015.

8 Hippocrates, *On Ulcers*, Part 7, trans. Francis Adams, www.classics.mit.edu, accessed 8 September 2015.

9 Peter Marshall, *The Magic Circle of Rudolf II: Alchemy and Astrology in Renaissance Prague* (New York, 2006), pp. 155–6.

10 L. Lewis Wall, 'The Medical Ethics of Dr J. Marion Sims: A Fresh Look at the Historical Record', *Journal of Medical Ethics*, xxxii/6 (June 2006), pp. 346–50.

11 J. Marion Sims, *The Story of My Life* (New York, 1884), p. 245.

12 'Size of the Nanoscale', www.nano.gov, accessed 13 September 2015.

13 Nate Seltenrich, 'Nanosilver: Weighing the Risks and Benefits', *Environmental Health Perspectives*, cxxi/7 (July 2013), www.ehp.niehs.nih.gov, accessed 15 September 2015.

14 Samuel Luoma, *Silver Nanotechnologies and the Environment: Old Problems or New Challenges* (Washington, DC, 2008), p. 5.

15 Seltenrich, 'Nanosilver: Weighing the Risks and Benefits'.

16 USGS, *The Lifecycle of Silver*, p. 6.

17 The Silver Institute, *World Silver Survey 2015: A Summary* (Washington, DC, 2015), p. 7.

18 The Silver Institute, 'Solar Energy', www.silverinstitute.org, accessed 18 September 2015.

7 Status Symbols

1 Meert Katrien, Mario Pandelaere and Vanessa M. Patrick, 'Taking a Shine to It: How the Preference for Glossy Stems from an Innate Need for Water', *Journal of Consumer Psychology*, XXIV/2 (2014), pp. 195–206.

2 N. H. Gale and Z. A. Stos-Gale, 'Ancient Egyptian Silver', *Journal of Egyptian Archaeology*, LXVII (1981), pp. 108–9.

3 Asian Art Museum of San Francisco, *Tomb Treasures from China: The Buried Art of Ancient Xi'an* (San Francisco, CA, 1994), p. 17.

4 Ruth E. Leader-Newby, *Silver and Society in Late Antiquity* (Aldershot, 2004), p. 1.

5 Peter Landesman, 'The Curse of the Sevso Silver', *The Atlantic* (November 2001), pp. 63–90.

6 Ilaria Gozzini Giacosa, *A Taste of Ancient Rome*, trans. Anna Herklotz (Chicago, IL, 1994).

7 Tracey Albainy, 'Eighteenth-century French Silver in the Elizabeth Parke Firestone Collection', *Bulletin of the Detroit Institute of Arts*, LXXIII/1–2 (1999), p. 13.

8 Peter Fuhring, 'The Silver Inventory from 1741 of Louis, Duc d'Orleans', *Cleveland Studies in the History of Art*, VIII (2003), p. 35.

9 Reproduced in Jean-Louis Flandrin, *Arranging the Meal: A History of Table Service in France*, trans. Julie E. Johnson (Berkeley, CA, 2007), p. 8.

10 Quoted ibid., p. 9.

11 Anonymous, *The Manners and Tone of Good Society* (London, 1880), p. 78.

12 Serena Bechtel, 'Changing Perceptions of Children, *c.* 1850–*c.* 1925, as Reflected in American Silver', *Studies in the Decorative Arts*, VI/2 (Spring–Summer 1999), pp. 84–5.

13 Thorstein Veblen, *The Theory of the Leisure Class* (New York, 1899), kindle location 1348.

14 Susan Mosher Stuard, *Gilding the Market: Luxury and Fashion in Fourteenth-century Italy* (Philadelphia, PA, 2006), p. 1.

15 Ibid., p. 49.

16 Valery M. Garrett, *Chinese Dragon Robes* (Oxford, 1998), pp. 1–5.

17 Quoted in Lucy Worsley, *The Courtiers* (London, 2010), p. 20.

8 **Pure**

1 Stephen King, *Cycle of the Werewolf* (New York, 1985).

2 Matthew Beresford, *From Demons to Dracula: The Creation of the Modern Vampire Myth* (London, 2008), p. 42.

3 Sabiha Al Khemir, *Nur: Light in Art and Science from the Islamic World* (Seville, 2014), pp. 14–22.

4 Trevor Bryce, *The Kingdom of the Hittites* (Oxford, 1998), p. 74.

5 Mathilde Avisseau-Broustet, Cécile Colonna and Kenneth Lapatin, 'The Berthouville Treasure: A Discovery "As Marvelous as It Was Unexpected"', in *The Berthouville Silver Treasure and Roman Luxury*, ed. Kenneth Lapatin (Los Angeles, CA, 2014), pp. 17–20.

6 Ruth E. Leader, *Silver and Society in Late Antiquity* (Aldershot, 2004), pp. 6–7.

7 Marlia Mundell Mango, *Silver from Early Byzantium: The Kaper Koraon and Related Treasures* (Baltimore, MD, 1986), p. 3.

8 Victor Elbern, 'Altar Implements and Liturgical Objects', in *Age of Spirituality: Late Antique and Early Christian Art, Third to Seventh Century*, ed. Kurt Weitzmann (New York, 1978), p. 592.

9 'Silver Plaque and Gold Disc from the Water Newton Treasure', www.britishmuseum.org, accessed 27 September 2015.

10 Quoted in J.N.D. Kelly, *Golden Mouth: The Story of John Chrysostom, Ascetic, Preacher, Bishop* (Grand Rapids, MI, 1995), p. 136.

11 Ruth E. Leader, *Silver and Society*, p. 67.

12 John Chrysostom, *On Wealth and Poverty*, trans. Catharine P. Roth (Crestwood, NY, 1984), p. 117.

13 Mu Soeng, *The Diamond Sutra: Transforming the Way We See the World* (Boston, MA, 2000), p. 62.

14 Sally Wriggins, *The Silk Road Journey with Xuanzang* (New York, 2004), p. 116.

15 Vesna Wallace, 'Mongolian Buddhist Manuscript Culture', *Buddhist Manuscript Cultures: Knowledge, Ritual and Art*, ed. Stephen C. Berkwitz, Juliane Schober and Claudia Brown (New York, 2009), pp. 83–4.

16 Hadith quoted in *God Is Beautiful and Loves Beauty: The Object in Islamic Art and Culture*, ed. Sheila Blair and Jonathan Bloom (New Haven, CT, 2013).

17 Tara E. Nummedal, *Alchemy and Authority in the Holy Roman Empire* (Chicago, IL, 2007), pp. 87–90.

18 Peter Marshall, *The Magic Circle of Rudolf II: Alchemy and Astrology in Renaissance Prague* (New York, 2006), p. 128.

19 Mircea Eliade, *The Forge and the Crucible: The Origins and Structure of Alchemy* (Chicago, IL, 1978), p. 159.

20 Carl Jung, *The Collected Works*, vol. XII: *Psychology and Alchemy*,
    ed. Herbert Read, trans. R.F.C. Hull (New York, 1968), p. 232.
21 'Evil in the Hebrew Bible, Mishnah, and Talmud', *Explaining Evil*,
    vol. I, ed. J. Harold Ellens (Oxford, 2011), p. 154.
22 Laurann Gilbertson, 'To Ward Off Evil: Metal on Norwegian Folk
    Dress', in *Folk Dress in Europe and Anatolia: Beliefs about Protection
    and Fertility*, ed. Linda Welters (Oxford, 1999), p. 201.
23 Ibid., p. 205.
24 *The Mysteries of Purity*, trans. Nabih Amin Faris (Lahore, 1991), p. 2.
25 See Virginia Smith, *Clean: A History of Personal Hygiene and Purity*
    (Oxford, 2007).

SELECT BIBLIOGRAPHY

Agricola, Georgius, *De re metallica* (1556), trans. Herbert Clark Hoover
    and Lou Henry Hoover (1912), www.gutenberg.org, accessed
    5 July 2016
Alcorn, Ellenor M., *English Silver in the Museum of Fine Arts, Boston*,
    vol. I (Boston, MA, 1993); vol. II (2000)
Blair, Claude, *The History of Silver* (New York, 1987)
Blair, Sheila, and Jonathan Bloom, eds, *God Is Beautiful and Loves
    Beauty: The Object in Islamic Art and Culture* (New Haven, CT, 2013)
Blainey, Geoffrey, *The Rush That Never Ended: A History of Australian
    Mining* (Melbourne, 1963)
Brown, Kendall, *A History of Mining in Latin America from the Colonial
    Era to the Present* (Albuquerque, NM, 2012)
Browne, John, *Seven Elements That Changed the World: An Adventure
    of Ingenuity and Discovery* (New York, 2014)
Cirillo, Dexter, *Southwestern Indian Jewelry* (New York, 1992)
Eliade, Mircea, *The Forge and the Crucible: The Origins and Structures
    of Alchemy* (New York, 1962)
Ferguson, Niall, *The Ascent of Money: A Financial History of the World*
    (New York, 2008)
Flandrin, Jean-Louis, *Arranging the Meal: A History of Table Service in
    France*, trans. Julie E. Johnson (Berkeley, CA, 2007)
Flynn, Dennis Owen, Arturo Giráldez, and Richard von Glahn, *Global
    Connections and Monetary History, 1470–1800* (Aldershot, 2003)
Garrett, Valery M., *Chinese Dragon Robes* (Oxford, 1998)
Gozzini Giacosa, Ilaria, and Anna Herklotz, *A Taste of Ancient Rome*
    (Chicago, IL, 1994)
Giráldez, Arturo, *The Age of Trade: The Manila Galleons and the Dawn
    of the Global Economy* (Lanham, MD, 2015)
Glanville, Philippa, *Silver in Tudor and Early Stuart England:
    A Social History and Catalogue of the National Collection, 1480–1660*
    (London, 1990)

—, and Jennifer F. Goldsborough, *Women Silversmiths, 1685–1845: Works from the Collection of the National Museum of Women in the Arts* (London and Washington, DC, 1990)

Hornblower, Simon, *The Greek World, 479–323 BC* (London, 1983)

Kamen, Henry, *Golden Age Spain* (London, 2005)

Kean, Sam, *The Disappearing Spoon: And Other True Tales of Madness, Love, and the History of the World from the Periodic Table of the Elements* (New York, 2010)

Khemir, Sabiha, *Nur: Light in Art and Science from the Islamic World* (Seville, 2014)

Lapatin, Kenneth, ed., *The Berthouville Silver Treasure and Roman Luxury* (Los Angeles, CA, 2014)

Leader-Newby, Ruth E., *Silver and Society in Late Antiquity* (Aldershot, 2004)

Lingenfelter, Richard, *Bonanzas and Borrascas: Gold Lust and Silver Sharks, 1848–1884* (Norman, OK, 2012)

Ma, John, Nikolaos Papazarkadas and Robert Parker, eds, *Interpreting the Athenian Empire* (London, 2009)

MacGregor, Neil, *A History of the World in 100 Objects* (London, 2010)

Maltby, William S., *The Rise and Fall of the Spanish Empire* (New York, 2009)

Mann, Charles C., *1493: Uncovering the New World Columbus Created* (New York, 2011)

Marshall, Peter, *The Magic Circle of Rudolf II: Alchemy and Astrology in Renaissance Prague* (New York, 2006)

Merriman, Philippa, *Silver* (London, 2009)

Minick, Scott, and Ping Jiao, *Arts and Crafts of China [Zhongguo gong yi mei shu]* (New York, 1996)

Miodownik, Mark, *Stuff Matters: The Strange Stories of the Marvellous Materials That Shape our Man-made World* (London, 2013)

Moreton, Stephen, *Bonanzas and Jacobites: The Story of the Silver Glen* (Edinburgh, 2007)

Nummedal, Tara E., *Alchemy and Authority in the Holy Roman Empire* (Chicago, IL, 2007)

Schroder, Timothy, *The National Trust Book of English Domestic Silver, 1500–1900* (New York, 1988)

Sims, J. Marion, *The Story of My Life* (New York, 1884)

Stevenson, Robert Louis, *The Silverado Squatters* (London, 1883)

Stuard, Susan Mosher, *Gilding the Market: Luxury and Fashion in Fourteenth-century Italy* (Philadelphia, PA, 2006)

Thomas, Hugh, *World Without End: The Global Empire of Philip II* (London, 2014)

Veblen, Thorstein, *The Theory of the Leisure Class* (New York, 1899)

von Glahn, Richard, *Fountain of Fortune: Money and Monetary Policy in China, 1000–1700* (Berkeley, CA, 1996)
Worsley, Lucy, *The Courtiers* (London, 2010)

ASSOCIATIONS AND WEBSITES

American Numismatic Society
www.numismatics.org

Australian Atlas of Minerals, Resources, Mines and Processing Centres
www.australianminesatlas.gov.au

Chinese Export Silver
chinese-export-silver.com

The Goldsmiths' Company
www.thegoldsmiths.co.uk

The J. Paul Getty Museum photography and silver collections
www.getty.edu/art/collection

Makers' Marks on British and Irish Silver
www.silvermakersmarks.co.uk

Mindat
www.mindat.org

Mining History Association
www.mininghistoryassociation.org

The National Mining Association
www.nma.org

Online Encyclopedia of Silver Marks, Hallmarks and Makers' Marks
www.925-1000.com

The Royal Mint
www.royalmint.com

Royal Society of Chemistry
www.rsc.org

The Silver Institute
www.silverinstitute.org

Society of American Silversmiths
www.silversmithing.com

United States Geological Survey (USGS) Minerals Information: Silver
minerals.usgs.gov/minerals/pubs/commodity/silver

The Victoria and Albert Museum
The museum's silver galleries can be searched online
www.vam.ac.uk

ACKNOWLEDGEMENTS

At times, writing about silver felt like writing about the history of the world, and so I am especially grateful to all the people who patiently discussed the intricacies of their fields. I would like to thank physicist Bruce Bray, chemists Eddie Moler, Stephen Moreton and Byron Shen, and geologists Alan Gallegos of the Sequoia National Park, Dana Willis of Coeur Mining, Charles Straw of Silver Mines, Ltd, Sydney, and Mike Huggins.

I was fortunate to receive help from many museum professionals in the United Kingdom, United States and China. I am grateful to all the curators and collections managers whose paths I crossed over the course of writing this book. I would like to give special thanks to Laura Belani of the Bowers Museum, Santa Ana, California; Laurann Gilbertson of the Vesterheim Norwegian-American Museum, Decorah, Iowa; the Shanghai Bank Museum; and Richard Sieber of the Philadelphia Museum of Art.

I also owe a debt to the many librarians and archivists who assisted me in my research, especially the staff at the Library of the Canyons, Silverado, California; the staff of the Bibliotheca Zi-Ka-Wei, Shanghai; the librarians at the library of the Royal Asiatic Society, China in Shanghai. I would also like to thank Elena Stolyarik of the American Numismatic Society.

Silver is not magnetic, but silversmiths often are. I would like to thank all the silversmiths, artists and jewellers who shared their passion and knowledge with me. Silversmith Charles Hall was especially generous with his time. I would also like to thank Anastasia Azure, Anthony Elson and Mielle Harvey. The painter Leslie Lewis Sigler provided stimulating conversation about silver and family heirlooms. The numerous collectors who offered assistance included lifelong friend Rodney Schwartz; Chris Buckley, an indefatigable source of knowledge on Chinese design; and Eric Boudot, whose knowledge of southwestern Chinese silver was most helpful.

Xiaolan Liu of Brocade Country, Shanghai, was a generous and patient source of information on Miao silver. I am also grateful to Michael

Garland of Garland's Jewelry, Sedona, Arizona; Tal Arshi of Yemenite Art, Jaffa, Israel; Christina Jansen of The Scottish Gallery, Edinburgh; Kathleen Slater of Adrian Sassoon, London; Beverly Bueninck of Christie's, Kevin Ward of Exceptional Minerals; and David Murray of Leopard Antiques. Michael DiRienzo, Director of the Silver Institute, Washington, DC, provided helpful industry contacts.

At Reaktion Books, I am especially grateful to Michael Leaman for his encouragement, Amy Salter for her patient and diligent editing and the multi-talented Harry Gilonis, picture editor and all-round inspiration.

Finally, I would like to thank my family: my parents Ewan and Audrey Macbeth for, among so many other things, their gifts of silver throughout my life, and also my husband Byron Shen and sons Ewan and Owen Shen for allowing me the space to write *Silver*.

PHOTO ACKNOWLEDGEMENTS

The author and publishers wish to express their thanks to the below sources of illustrative material and/or permission to reproduce it. (Some information not placed in the captions for reasons of brevity is also given below.)

From Georgius Agricola, *De re metallica libri XII* (Basel, 1556), photo Beinecke Rare Book and Manuscript Library, Yale University, New Haven, CT: p. 36; image courtesy American Numismatic Society: p. 91; photos author: pp. 48, 52, 67; Bayerisches Nationalmuseum, Munich: p. 135; courtesy Ben-Zion David, Yemenite Art: p. 66; courtesy the Bowers Museum, Santa Ana: pp. 52, 50; The British Museum, London (photo © Trustees of the British Museum): p. 181; images courtesy Chris Buckley: pp. 172, 173; image © Christie's Images: p. 140; courtesy Coeur Mining: pp. 22, 24, 55; Photo FA2010: p. 135; photo GanMed64: p. 57 (this file is licensed under the Creative Commons Attribution 2.0 Generic license: any reader is free to share – to copy, distribute and transmit the work, or to remix – to adapt the work, under the following conditions: you must attribute the work in the manner specified by the author or licensor [but not in any way that suggests that they endorse you or your use of the work] – you may not apply legal terms or technological measures that legally restrict others from doing anything the license permits); image courtesy Geographicus Rare Antique Maps, provided to Wikimedia Commons: pp. 110–11; image courtesy The Goldsmiths' Company (photographer Richard Valencia): p. 169; courtesy Charles Hall: pp. 6, 8, 10; courtesy Mielle Harvey: p. 63 (foot); courtesy Mike Huggins: p. 50; The J. Paul Getty Museum, California: pp. 61, 62, 69 foot, 77, 82, 119, 159, 162 (images courtesy of the Getty's Open Content Program); Jewish Museum, New York: p. 168; photo Paco naranjo jimenez: pp. 34–5 (this file is licensed under the Creative Commons Attribution-Share Alike 4.0 International license: any reader is free to share – to copy, distribute and transmit the work, or to remix – to adapt the work, under the following conditions: you must attribute the work in the manner specified by the author or licensor (but not in any way that suggests that

INDEX